Graph Theoretical Models of Abstract Musical Transformation

Recent Titles in
the Music Reference Collection

A Guide to Popular Music Reference Books: An Annotated Bibliography
Gary Haggerty

Rock Music Scholarship: An Interdisciplinary Bibliography
Jeffrey N. Gatten

Sing Glory and Hallelujah!: Historical and Biographical Guide to *Gospel Hymns Nos. 1 to 6 Complete*
Samuel J. Rogal, compiler

Orchestration Theory: A Bibliography
James E. Perone, compiler

Brass Music of Black Composers: A Bibliography
Aaron Horne, compiler

The Music and Dance of the World's Religions: A Comprehensive, Annotated Bibliography of Materials in the English Language
E. Gardner Rust

Twentieth-Century American Music for the Dance: A Bibliography
Isabelle Emerson, compiler and editor

Electronic and Computer Music
Robert L. Wick

Harmony Theory: A Bibliography
James E. Perone, compiler

A Mozart Diary: A Chronological Reconstruction of the Composer's Life, 1761–1791
Peter Dimond, compiler

The Grateful Dead and the Deadheads: An Annotated Bibliography
David G. Dodd and Robert G. Weiner

A Sourcebook of Nineteenth-Century American Sacred Music for Brass Instruments
Mark J. Anderson

Graph Theoretical Models of Abstract Musical Transformation

An Introduction and Compendium for Composers and Theorists

Jeffrey Johnson

Music Reference Collection, Number 62

Greenwood Press
Westport, Connecticut • London

Library of Congress Cataloging-in-Publication Data

Johnson, Jeffrey.
Graph theoretical models of abstract musical transformation : an introduction and compendium for composers and theorists / Jeffrey Johnson.
p. cm.—(Music reference collection, ISSN 0736–7740 ; no. 62)
Includes bibliographical references.
ISBN 0–313–30158–1 (alk. paper)
1. Music—Theory—20th century. 2. Graph theory. 3. Composition (Music) 4. Musical analysis. I. Title. II. Series.
MT6.J65G7 1997
781—DC21 97–2739
MN

British Library Cataloguing in Publication Data is available.

Library of Congress Catalog Card Number: 97–2739
ISBN: 0–313–30158–1
ISSN: 0736–7740

First published in 1997

Greenwood Press, 88 Post Road West, Westport, CT 06881
An imprint of Greenwood Publishing Group, Inc.

Printed in the United States of America

The paper used in this book complies with the Permanent Paper Standard issued by the National Information Standards Organization (Z39.48–1984).

10 9 8 7 6 5 4 3 2 1

Copyright Acknowledgments

Contents

Preface

Debate may have arisen, in a different age, whether the study of abstract musical transformation should be classified as musica speculativa. After all, it is a study that collects resonances from many disciplines and often produces diagrams of aesthetic if not mystic significance. But unlike its cosmic and occult relatives in musica speculativa, graphs of abstract musical transformations flow very naturally into practical application as an analytical tool, or as a means to inspire and generate compositional thought.

The background against which abstract musical transformations appear is referred to as a musical space. The conception and description of musical spaces, and transformations that exist within them, changed radically with the publication of *Generalized Musical Intervals and Transformations* by David Lewin in 1987. This work identified the similarity between entities and operations by introducing generalized interval systems (or GIS), and formalized fundamental constructs such as transformation graphs and transformation networks. The abstract musical transformations modeled in this book are an intensification of graph-theoretical aspects of Lewin transformation networks.

The most obvious, accessible, and perhaps ultimately most practicable abstract musical transformations involve the creation and manipulation of pitch-class networks. These networks, in which each vertex is assigned a distinct pitch-class, create a system of relationships vast and powerful. Extending beyond them, any self-contained system of musical parameters, instructions, or quotations can be modeled by diagrams displayed in this book, which should be understood in this sense as an introduction meant to inspire exploration of possibilities.

Graphic diagrams appeared early in musical literature with functions ranging from illustrative to comparative. Graph-theoretical applications differ, in that the actual diagram itself is perhaps the least important piece of information conveyed, compelling and suggestive as it may be. Graph theory allows us to construct and penetrate deeper levels of connection between objects, or to create objects by manipulating basic principles. The distinction between Cartesian and non-Cartesian graphing is central to understanding abstract musical transformation, and applications of both from theoretical literature will be examined.

In analytical applications, an understanding of graph-theoretical models of abstract musical transformation can be used descriptively to display ordered patterning among pitch-classes, or other objects in a musical composition. Applied to more substantial high-level explanatory endeavors, they can show structural correlation among various levels of parameters by providing a common notation for pitch, rhythmic, sectional or any other discrete musical parameter.

Although knowledge of abstract musical transformation can be valuable in many contexts as an analytical mechanism, it seems to have even greater potential for compositional applications. The use of these diagrams to generate compositional surfaces differs significantly from their use in analysis: an analytical model relates to a single musical surface, whereas compositional applications can be used to generate any potential surface derived from construction of the graphs. One needs only the curiosity necessary to transform abstract ideas into compositional practice. Then, all that remains is a sense for how characteristics of a transformation graph can influence and inspire compositional surfaces, and how to pursue systematically designs that resonate with all phases of the compositional imagination.

Prior training in, and experience with, basic notational practices of music-theoretical literature is assumed. This book is based upon insight derived from a community of theorists and composers, and without that foundation it could not have been conceived. It also was influenced and shaped by insight derived from members of the mathematical community, including most importantly Paul Erdös, Frank Harary, and Richard J. Trudeau.

This book is organized into four parts, the first introducing the musical-historical framework into which this study fits, and establishing formalities, essential terminology, and techniques. Part two develops constructions of graphic figures. The third part establishes a classification scheme, not only for organizational purposes, but also to help in guiding the selection of designs appropriate to various compositional plans. Part four suggests types of musical parameters or passages that can be assigned to the vertices of a transformation network. It also examines the literal spaces associated with some internal closures and external transformation structures. It concludes with an analysis using pitch-class networks in a work by Stefan Wolpe.

The appendices present a collection of degree sequences most practical for musical application. All 156 nonisomorphic complementary hexachordal graphs are displayed in Appendix I. Appendix II lists all twelve-position degree sequences in descending order from 5-regular down to 2-regular. Appendix III lists twelve-position degree sequences by total number of edges required from eleven through seventeen total edges.

A database of the complete degree sequences on twelve vertices is available by FTP from ftp://ftp.prz.tu-berlin.de/pub/MusicAnalysis/Databases/ or from midiworld. http://midiworld.com/

Part One

GRAPHING

Introduction

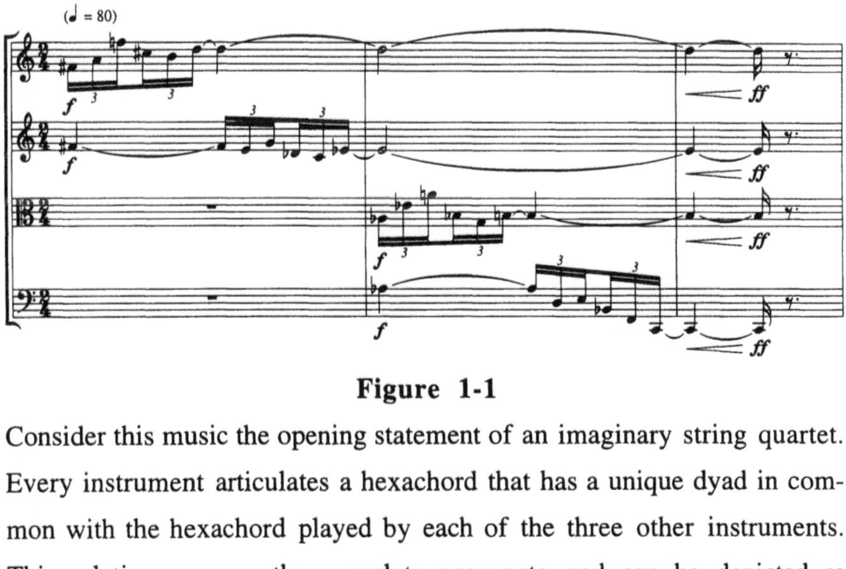

Figure 1-1

Consider this music the opening statement of an imaginary string quartet. Every instrument articulates a hexachord that has a unique dyad in common with the hexachord played by each of the three other instruments. This relation engages the complete aggregate and can be depicted as shown in Figure 1-2.

Each instrument of the quartet is represented by a circle in Figure 1-2,

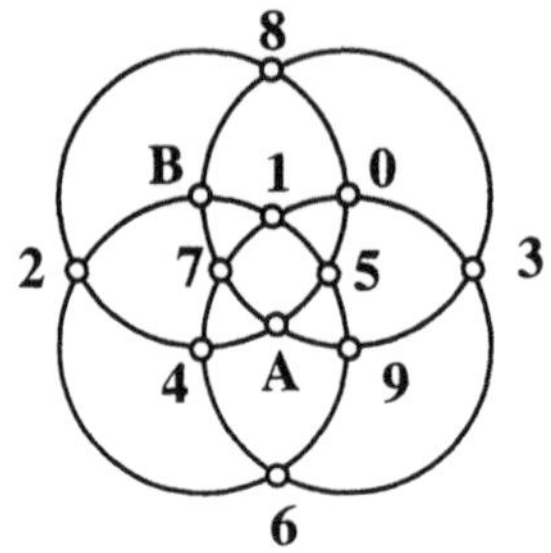

Figure 1-2

with every pitch-class* played by that instrument displayed in order around it. The music played by the first violin in Figure 1-1 is represented by counterclockwise movement around the lower-left circle beginning at pitch-class <6>, the same pitch-class that begins the second violin line, shown as clockwise motion around the lower-right circle. The viola line is represented by clockwise motion around the upper-right circle beginning at pitch-class <8>, and the cello by counterclockwise motion around upper-left circle, also beginning at pitch-class <8>. This representation of the musical surface is a graph.

Hexachordal interconnections in Figure 1-1 are visually clarified using a graph. Since each pitch-class is represented by a dot (called a vertex), and the relation of each pitch-class to other pitch-classes is represented by connecting lines (called edges), this diagram is a summary of pitch-class orderings in this passage.

In the continuation of our imaginary string quartet, relationships and relative positions of all pitch-classes in Figure 1-2 could remain constant. In addition to being a model of the opening passage, the diagram could be

---

* Pitch-class equivalence creates twelve distinct entities, each representing one of the twelve chromatic pitches within any octave. Numeric notation is used in this book to represent pitch-classes with the following scheme:

0	**1**	**2**	**3**	**4**	**5**	**6**	**7**	**8**	**9**	**A**	**B**
C	**C#**	**D**	**D#**	**E**	**F**	**F#**	**G**	**G#**	**A**	**A#**	**B**
	Db		**Eb**			**Gb**		**Ab**		**Bb**	

used to generate both large and small-scale pitch relations for the entire piece. Considered as a compositional generator, musical structures could be projected that develop the opening orderings by using different pathways through and around the graph. In Figure 1-3, the first violin moves clockwise around the outer tetrachord of Figure 1-2, while the second violin and viola move within the complementary octachord in various ways.

Figure 1-3

Extending beyond the first movement of our imaginary quartet, we could choose to develop the graph in subsequent movements by transforming, or simply rearranging the way in which pitch-classes interact on the graph itself. If the specific pitch-classes in Figure 1-2 are removed, the basic connective relations between vertices that remain become an abstract model, onto which pitch-classes or other musical objects can be reassigned. When pitch-classes are reassigned in various orderings, individualized profiles emerge with possibilities various and extreme. Each ordering produces unique characteristic intervallic patterns, but all share fundamental connective relations incorporated in abstract connective properties of the graph itself.

Descriptive graphs present single-level abstractions, but graphs of abstract musical transformations are multilevel in that each design has abstract characteristics and properties independent of those created by assigning elements such as pitch-classes, onto it. The distinction and generalization of these levels will use machinery developed from the second half of David Lewin's *Generalized Musical Intervals and Transformations*. In chapters seven through ten, Lewin provides formal definitions, musical examples, and applications for node/arrow systems, transformation graphs, and transformation networks. A node/arrow system is defined as a diagram with nodes as vertices and arrows as edges. To make a node/arrow system into a transformation graph, one labels all edges with self-consistent names of functions from within a subgroup. A transformation network is derived from a transformation graph by labeling nodes with objects in a manner consistent with the arrows.

"Node/arrow" terminology was no doubt chosen to avoid the informal baggage that comes with the word "graph." Successful where connection with the field of graph theory is not a principle focus, mechanisms from, and associations with the field of graph theory are an integral part of this book, so the term graph will be used instead of node/arrow system.

It is important to have a clear understanding of Lewin's distinction between transformation graph and transformation network. Self-consistent transformations are assigned to the edges of a graph to produce a transformation graph. These transformations can be twelve-tone operators such as transposition, inversion, or cycle of fifths multiplicative transforms. This transformational action can also be considered more generally, where edges of the graph represent a function mapping the contents of its vertex into the contents of an adjacent vertex. This more generalized transformational action will be called a circulation function. As soon as self-consistent circulation functions are in place a graph becomes a transformation graph.

With a transformation graph, relations among objects define the musical potential at least as much as, if not more than, the individualistic atomic elements themselves. It is this nonCartesian approach to relationships that makes abstract musical transformations so powerful. When pitch-classes are assigned onto the vertices of a transformation graph that resonate with it, the resultant diagram will be called a pitch-class network. Consider the hierarchy of this concept diagrammed in Figure 1-4.

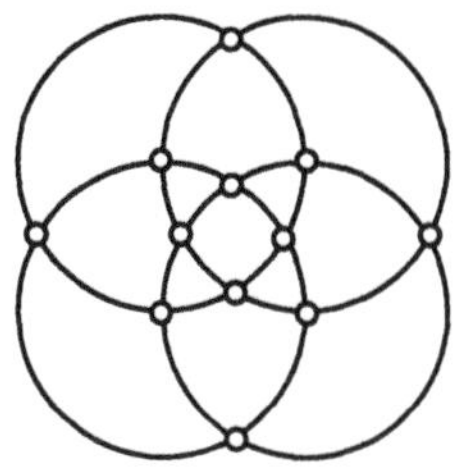

Figure 1-4A: Graph

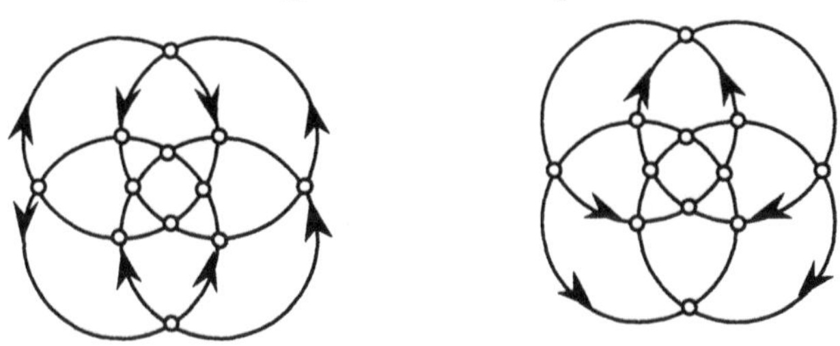

Figure 1-4B: Transformation Graphs developed from Figure 1-4A

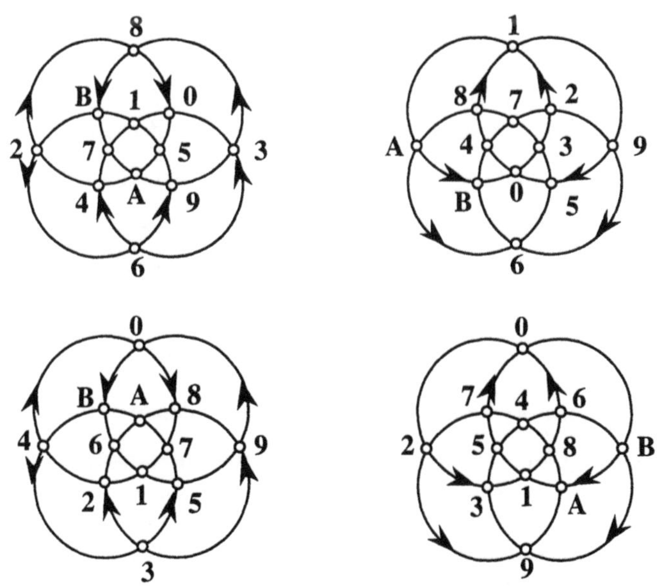

Figure 1-4C: Four pitch-class networks developed from Figure 1-4B

The inspiration for, and creation of connections, displayed by graphs occupies a distinct phase of the compositional process earlier than, and in many cases separate from, realization and arrangement of objects onto the diagrams. It engages a complex and deeply hierarchic aspect of precompositional planning during which creation and organization of a musical space takes place. In the article "Compositional spaces and other territories" (Morris, 1995C), Robert Morris has begun to chart this phase of the compositional process. A compositional space is a conceptual link in the creative process establishing a collection of musical objects connected by at least one principle; distinguishing this stage of planning from compositional design, where orderings are implemented. Morris describes them as "out-of-time structures from which the more specific and temporally oriented compositional design can be composed" (p.330). He defines two categories of compositional spaces, literal and abstract spaces. Literal spaces have a single pitch-class representation, whereas abstract spaces can generate several literal spaces.

Although fairly intuitive for many composers, the idea itself has been underexamined. Morris gives examples of Boulez multiplication matrices, generalized Mead tiles, and the "Babbitt network" of trichordal set-classes that partition all-combinatorial hexachords. He also cites other "abstract" spaces including Room squares, two-partition graphs, and musical grammars. Stefan Wolpe touches briefly upon the idea of compositional spaces in the article "Thinking twice" in hopes of bringing expansion to the "partly very mechanistic and neutral" quality he observed as a dangerous occupational hazard when using "the saturated balance of twelve tones."

> Generally, one understands the all-chromatic circuit as an equilibrium of tones whose centralizing effects result from consecutive realizations of complementary pitch relations. Through progressive summations of single pitch-strands, the center is extended throughout the compositional space. If the center is extended throughout the compositional space, then all the sectors in it, fed by their share of the center, become centers themselves. From this one can infer that

> the all-chromatic chain can be unhinged, its sections interrupted, isolated, and arrested. A constellatory space (to avoid the term "thematic space") will then be established in which the elements—the tones—are fashioned according to the degree of precompositional activity. (Wolpe, 1967: 277-8)

Iannis Xenakis uses the term "geometrizing" in *Formalized Music,* to help conceptualize the nature of the precompositional planning used during articulation of compositional spaces, and welcomes the application of a mathematical apparatus to help support its construction:

> The effort to make art while "geometrizing," [has] led to a sort of abstraction and formalization of the musical compositional act. This abstraction and formalization has found ... an unexpected and, I think, fertile support in certain areas of mathematics. It is not so much the inevitable use of mathematics that characterizes the attitude of these experiments, as in the overriding need to consider sound and music as a vast potential reservoir in which a knowledge of the laws of thought and the structured creations of thought may find a completely new medium of materialization, i.e., of communication. (Xenakis, 1971: ix)

The formalization and abstraction of the compositional space inhabited by abstract musical transformation has "fertile support" from the mathematical field of graph theory, to which, in return, it will offer a practical application of musical structures that can be created and described using its machinery. Graph-theoretical constructs will facilitate use of a self-contained dimension of compositional space occupied by abstract musical transformation. This continuum of structures comprises pitch-class networks: all subsets of the unordered collection of twelve pitch-classes, and models any other self-contained transformation structure, producing associations between audibly related or unrelated objects.

The remainder of this chapter will clarify essential distinctions between graph theory and graphs commonly encountered elsewhere. There will also be a brief discussion of the historical foundations of graph theory and the incorporation of graphs as music-analytical tools.

Musical Graphing

Graphs seem to surround us. From Wall Street trends to television advertising, developed from spreadsheets, passed out during business presentations, people generally appreciate the visual organization they afford. The most common are bar graphs and pie charts, where the diagram itself is a visual presentation of its own data that allows proportional comparison.

When intersecting vertical and horizontal axes are used to transmit changing parameters, a different type of graph arises. Graphs of algebraic equations are drawn in a planar space partitioning a Cartesian plane, upon which Cartesian graphs are created, named after René Descartes (1595-1650). Cartesian graphs have been historically useful to musicians, and many musical phenomena can readily be given graphic presentation.

Musical contour lends itself quite nicely to graphing because the discrete steps of a graph reveal motions in ways visually obscured by enharmonics and accidentals of our musical notational system. An early use of melodic contour graphs was proposed in Joseph Schillinger's first compositional method book (1941). With time on the abscissa (horizontal axis) and pitch on the ordinate (vertical axis), it presented a clear distinction between pitch and pitch-class, and related exact melodic shape to linear musical time.

Consider the opening of the second bagatelle from Béla Bartók's *Fourteen Bagatelles for Piano*, and a Cartesian graph of its melodic contour in Figure 1-5. The unfolding and dismantling of a symmetrical melodic structure balanced around A-440 becomes visually clear with this representation; much clearer than our notational system permits. It also reveals how each eighth rest in the left hand contributes to the larger pattern. This is an excellent method for depicting a composer's sense of timing, revealing sectional proportion in relation to the whole.

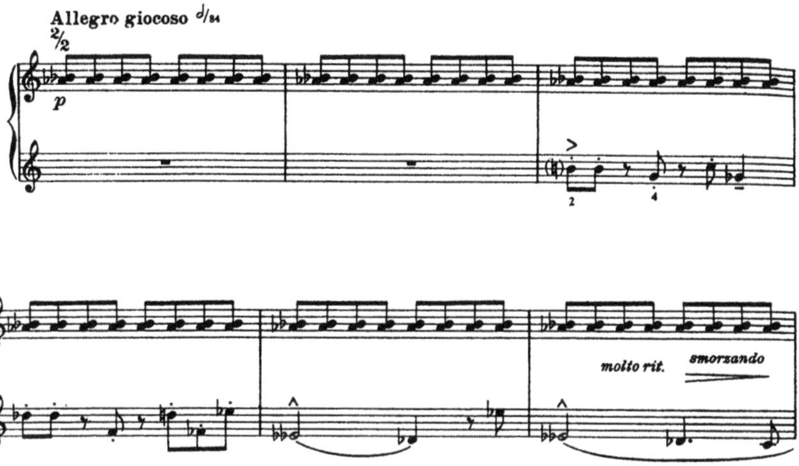
Allegro giocoso
p
molto rit.
smorzando

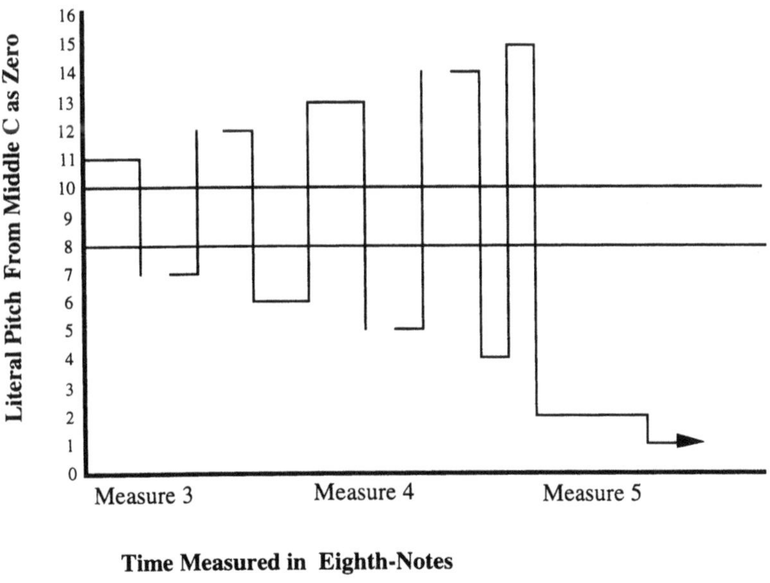
Literal Pitch From Middle C as Zero
16
15
14
13
12
11
10
9
8
7
6
5
4
3
2
1
0
Measure 3
Measure 4
Measure 5
Time Measured in Eighth-Notes

Figure 1-5

Schillinger also felt that graphic notation was more precise than standard notation, and should therefore replace it.

> Graph notation records individual components through individual curves. The special components of sound are frequency, intensity, and quality—and they may be recorded through the corresponding individual graphs. By means of such notation the composer can define his intentions with the utmost precision; the performer can then decipher the desires of the composer to the latter's full satisfaction. (Schillinger 1941:244.)

Performers are momentarily happy to hear this information until they read the next paragraph:

> In the future, with the elimination of the living performer, the graph method will still be valid for use with automatically performing musical instruments. Curves of composition and curves of execution will then merge into one. (Schillinger 1941:245.)

Other important and informative musical graphs can be found in *Sonic Design* by Robert Cogan and Pozzi Escot (1976), where acoustical graphs of large-scale motion as well as surface detail graphs are presented in diverse musical contexts. *Sonic Design* also broadened the application of graphing technique to include graphs of sound spectra.

As common and comfortable as they seem, a broader, much more generalized application radically different from common business perception of graphs has been developed by mathematicians, creating a branch of mathematics called graph theory. Graph-theoretical constructs are not dependent on Cartesian coordinates, but only on the abstract relationship between dots and connecting lines. Graph theory had its origin in solutions to puzzles and games rather than more intense desires to describe motion, quantify measurement, and, bring order to scientific inquiry that impelled other branches of mathematics. Its birth is generally accepted to have occurred with Leonhard Euler's paper from 1736 titled "The Seven Bridges of Königsberg." The question posed was whether a person could cross each of seven bridges between the city and a central island formed in the river, once, but not more than once. The solution to this famous problem began with representation of land masses as vertices, and bridges as edges of a graph. The mathematical proof determined that it was, in fact, impossible to cross each bridge only once. Euler's paper not only gave this proof, but gave a general method for solving other problems of the same type. Informally, dots represent any object or conception, and lines between dots represent a connection between them. According to the highly insightful book *Graph Theory 1736-1936* (Biggs, Lloyd, and Wilson, 1976), the term graph was first used in its modern sense by the chemist James Joseph Sylvester (1814-1897). Sylvester was infamous for inventing eccentric terminology, much of which has survived. In 1876, he came to the United States to accept a position as the first professor of mathematics at Johns Hopkins University. His historic invention of the term appears in the article "Chemistry and Algebra" (*Nature* 17:284, 1877-78). It is used in the context of the chemical graphic notation being developed then.

Graph theory has been used less in music-theoretical literature than has equation graphing, but where it has been applied the results have been engaging. James Bennighof (1987) used graphic displays in "Set-class aggregate structuring, graph theory, and some compositional strategies." With minimal formality he used graph theory to display the intervallic content of set-classes displayed as a cyclic mapping against the background structure of the aggregate. This technique makes an audible sensation visible and shows the relation of a part to its whole: a set-class as a pattern within the pitch-class aggregate. He represents this pattern by assigning each pitch-class to a dot (arbitrarily arranged in clockface order) and using lines to connect the total interval-class content of all transpositions and inversions of a single set-class.

In *Generalized Musical Intervals and Transformations,* David Lewin presents an argument for leaning away from Cartesian thinking toward more non-Cartesian processes that can create motions and contexts:

> We are very much under the influence of Cartesian thinking. We tend to conceive the primary objects in our musical spaces as atomic individual "elements" rather than contextually articulated phenomena like sets, melodic series, and the like. And we tend to imagine ourselves in the position of *observers* when we theorize about musical space; the space is "out there," away from our dancing bodies or singing voices. (Lewin 1987:158-59)

The thought process introduced requires a complement to the positivist approach to entities and operations. In the positivist approach, atomic musical events, which are most often pitch-classes, form larger collections of events such as sets and are classified as such. Lewin's approach defines atomic musical events as motion within a system defined by a network of transformations. Groupings that form set-classes are secondary to the context and transformations that produce them. Entities and operations are developed as two sides of the same principle. His use of graphing is not confined to pitch or pitch-class relations but incorporates any object (pitch, rhythm, or transformation itself) into a relationship of interlinked transformations.

The incorporation of formal graph theory into musical theoretic endeavor, compositional theory, and the newly developing ideas of compositional space and planning remains largely untapped, leaving potential for both explanatory and exploratory applications. Exploration of this vast area can take place quickly. It requires only a generalization of the elements comprising a graph, a means of detecting what can and cannot be graphed, a classification scheme of characteristics to analyze and select specific properties, and a method for creating visual displays. To use and develop ideas from this book, and for it to stimulate the curiosity necessary to transform abstract ideas into compositional practice, one needs to sense how characteristics of a transformation graph can influence and inspire compositional surfaces, and how to pursue designs systematically that resonate with all phases of the compositional imagination. This process must be rooted in a clear understanding of fundamental elements.

Formalities

A graph **G** can be created using a combination of three elements: a nonempty set of vertices (the formal name for dots), with an optional set of edges connecting vertices (the formal name for lines), and an incidence relation that associates the end of an edge with a vertex. For concise representation, these elements can be arranged in an ordered triple (V(G),E(G),IN(G)) where V(G) is a vertex set, E(G) is an edge set, and IN(G) is an incidence relation. As described by the elements of an ordered triple, no other form of representation is absolutely necessary, contrary to the common assumption that a graph must be pictorial. Pictorial representation does have many uses, however, one of which is to make interconnected relationships in an ordered triple understandable visually; each vertex is represented by a dot, each edge by a line, with lines connected to dots according to the incidence relation. Most graphs can be drawn in more than one way. All three graphs in Figure 1-6 look different, but the ordered triple of each would have the same contents, so they are considered equivalent. None of the drawings has special significance outside a context that would favor one as a more appropriate model than the other.

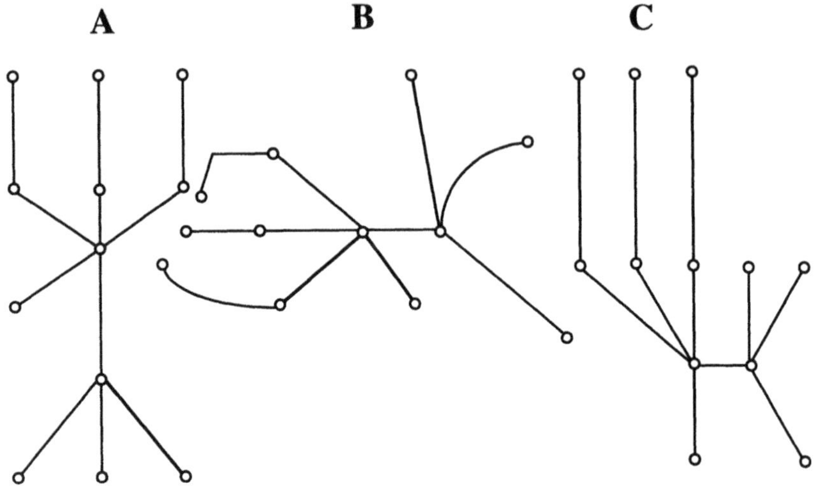

Figure 1-6 Three Equivalent Graphs

Vertices are connected if there is a single edge between them, and all connected vertices of a graph form connected components. If a diagram has more than one component, like Figure 1-7, then it is a system.

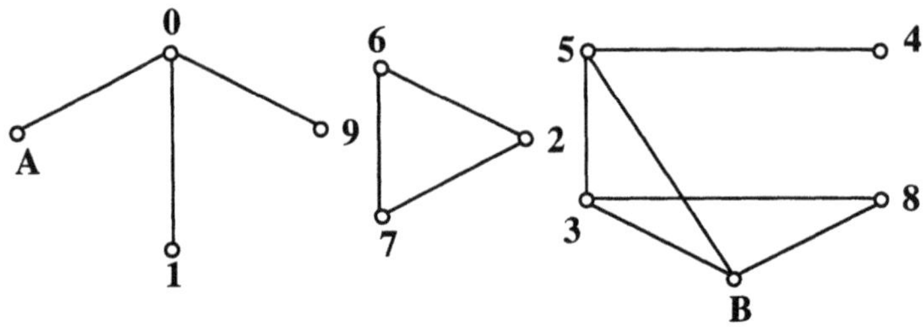

Figure 1-7 Three Connected Components in a System

In a system, each connected component forms an equivalence relation on the total vertex set. Separate components of a system could be distinguished by timbre, register, or temporal displacement. Whether a diagram contains a single connected component or a system, the first step in understanding its compositional potential is accomplished by developing the machinery to analyze each connected component as an independent unit, then afterwards to gauge the effect of its interaction within the system into which it would be placed. For pitch-class graphs, the number of distinct vertices in each connected component provides an obvious but useful criterion in sorting connected components. The number can range from one to twelve, each number of vertices having a finite and listable number of transformation graphs associated with it.

To be considered an edge, a line has to connect at least two vertices. Two vertices with an edge connecting them will be called adjacent vertices. Edges connecting the same vertex will be called adjacent edges. Edges are allowed to cross freely, without producing an additional vertex at the intersection unless so desired.

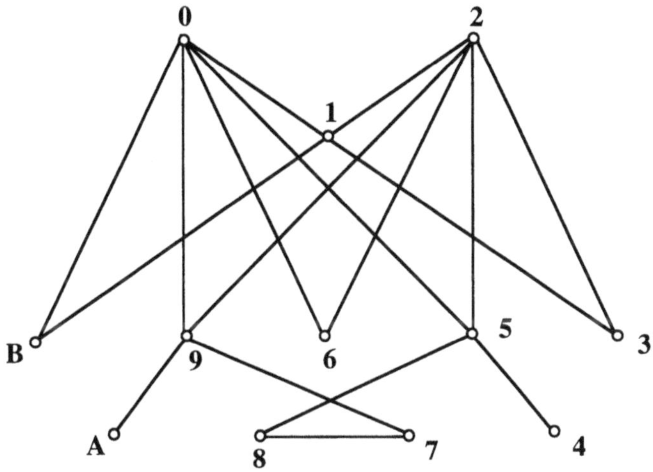

Figure 1-8

In the pitch-class network shown in Figure 1-8, pitch-class <2> can be transformed into pitch-class <6> directly without creating any vertices when it crosses the edges connecting pitch-classes <13> or <05>.

There are instances where connection between adjacent vertices is restricted or not permitted at all. It is therefore important to distinguish between adjacency and incidence. Vertices are adjacent if there is an edge between them; vertices are incident if connection can take place between them. There are two classes of connective motion possible on any edge. The pitch-classes <A9> are adjacent vertices in Figure 1-8.

A motion connecting only from <A> to <9>, **or** from <9> to <A> would be called restricted motion. Motion connecting <A> to <9> **and** <9> to <A> would be called unrestricted motion. Restricted motions cause incidence and adjacency to be unequal. Unrestricted motions allow adjacency and incidence to be equal. To prevent diagrams from becoming cluttered, arrows will only be used when direction is restricted in one direction, or where double arrows will help to clarify the diagram. Edges without arrows are implicitly to be considered unrestricted. If all edges in Figure 1-8 are considered unrestricted, then:

<0> is incident to 1,5,6,9,B
<1> is incident to 0,2,3,B
<2> is incident to 1,3,5,6,9
<3> is incident to 1,2
<4> is incident to 5
<5> is incident to 0,2,4,8
<6> is incident to 0,2
<7> is incident to 8,9
<8> is incident to 5,7
<9> is incident to 0,2,7,A
<A> is incident to 9
<B> is incident to 0,1

Restricted motion is implemented using arrows, altering incidence patterns. Four restricted motions alter Figure 1-8 to form Figure 1-9. In Figure 1-9:

<0> is incident to 1,5,6,9,B
<1> is incident to 0,2,3,B
<2> is incident to 1,3,5,9
<3> is incident to 1
<4> is incident to 5
<5> is incident to 0,2,4,8
<6> is incident to 2
<7> is incident to 8,9
<8> is incident to 7
<9> is incident to 0,2,7,A
<A> is incident to 9
<B> is incident to 0,1

In Figure 1-9, incidence TO is altered for pitch-classes <2>, <3>, <6>, and <8>. Correspondence between incidence FROM and incidence TO does not always exist in diagrams having restricted motions; notice that <0> is incident TO five vertices, but incident FROM only four. One can get to <0> in more ways than one can leave it. <0> is adjacent to <6> but is not incident FROM it.

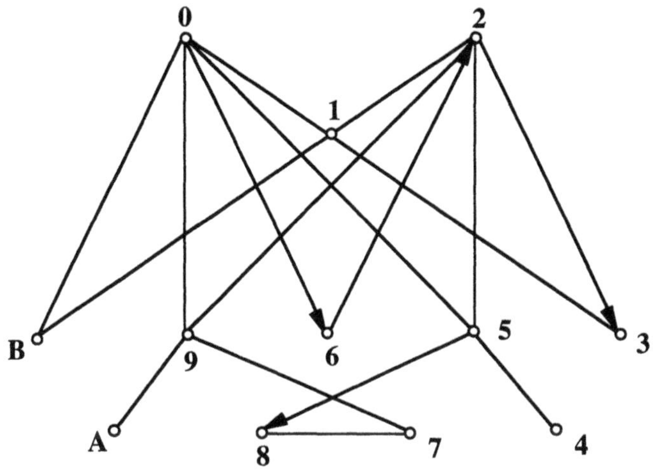

Figure 1-9

Compositional application of transformation networks involves planning motion from vertex to vertex in a series of patterns called tracks, forming material for interconnected musical surfaces. There are two basic track patterns: disjunctive tracks, having distinct end vertices, and cyclic tracks, with identical end vertices. In Figure 1-9, the track <0131258> is disjunct, <0B10> is a cycle. Disjunct tracks and cycles make up the most elemental realization actions.

Enumeration: Degree Sequences

The ordered triple, (V(G),E(G),IN(G)), is a nonpictorial representation of a graph. Another nonpictorial representation is a degree sequence: an organized listing of incidence in a graph, organized by the number of edges adjacent to each vertex. This format is concise and allows all possible graphs to be enumerated and collected systematically, without commitment to an exclusive visual representation for any of them.

The degree sequence of a graph is found by counting the edges incident to each vertex, and listing all of them in descending order. In the right-most component of Figure 1-7, vertex <4> has one edge; we say it is 1-incident. Vertices <3>, <5>, and <B> are 3-incident and vertex <8> is 2-incident. The degree sequence of the right-most component of Figure 1-7 is written **[33321]**.

To list all possible degree sequences for a given number of vertices, we could start by generating a list of all descending sequences for those vertices. For four vertices that list is:

[3333]	**[3322]**	**[3222]**	**[3111]**	**[2211]**
[3332]	**[3321]**	**[3221]**	**[2222]**	**[2111]**
[3331]	**[3311]**	**[3211]**	**[2221]**	**[1111]**

Some of these sequences cannot be made into degree sequences. If there is an odd number of odd vertices, the resulting diagram cannot be realized. Consider the degree sequence **[3222]**.

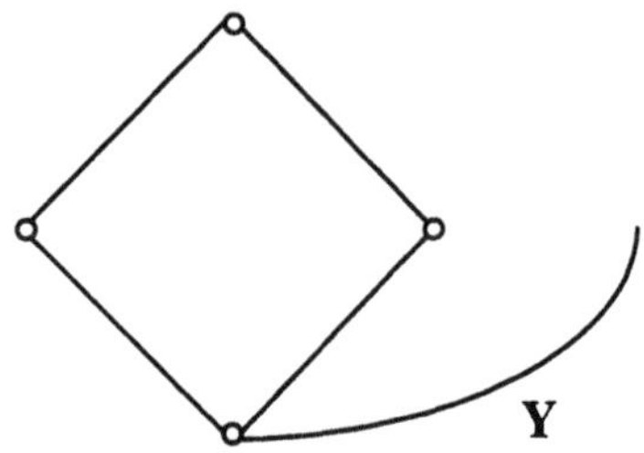

Figure 1-10

Notice that the edge marked Y cannot be connected without incrementing one of the three remaining vertices; and that would alter the desired degree sequence. Edge Y is referred to as a nonadjacent edge, and any graph with one or more nonadjacent edges will be called a nongraphic sequence. In eliminating all sequences with an odd number of odd vertices from our previous list the following sequences remain:

[3333]	**[3322]**	**[3221]**	**[2222]**	**[1111]**
[3331]	**[3311]**	**[3111]**	**[2211]**	

It is possible to construct all sequences in this list, but [1111] does not form a single connected component. The union of two disjoint components, [1111] is a system.

A given degree sequence must be a system if the summation of every element in its sequence is less than (2v-2) where v is the total number of vertices in the matrix. One last condition is necessary: degree sequences must produce relationships with no multiple edges between any two vertices. A multiple edge can be found connecting vertices <78> page 72. Multiple edges can be differentiated by timbre, register, or other parameters, and therefore can be made musically valid. A transformation graph with multiple edges will be termed complex, and with no double edges will be called simple. Complex graphs can be developed from simple graphs through addition of edges, therefore, complex graphs have a simple root; found by subtracting appropriate edges. For this reason, the enumeration of simple graphs is imperative.

An algorithm for enumerating all simple graphs was introduced by Erdös and Gallai (1960), and is based on a system of inequalities. This system is a series of comparisons between digit sums and the cumulative positional values within a degree sequence.

The Erdös and Gallai System of Inequalities

The theorem reads as follows:

A nonincreasing sequence of nonnegative integers S: $d_1, d_2, \ldots, d_p$ is graphical if, and only if,

$$\sum_{i=1}^{p} d_i \text{ is even}$$

(First Clause)

and for each integer k, $1 \leq k \leq p-1$,

$$\sum_{i=1}^{k} d_i \leq k(k-1) + \sum_{i=k+1}^{p} \min\{k, d_i\}.$$

(Second Clause)

An English translation of this theorem begins by arranging the degree sequence in nonincreasing order, so that numbers are either equal or descending when read left to right. The first clause is a simple test; the sequence is nongraphable if the sum of its degrees is an odd number.

In the second clause, a string of comparisons must be made between two accumulators; the one on the left-hand side of $\leq$ is compared to the one on the right-hand side. At any point in the second clause where the right-hand side is smaller than the left-hand side the sequence is proven to be nongraphable.

The variable **p** is the number of vertices in the sequence. The maximum number of stages in the comparison string will be one less than **p**. The variable **k** is a pointer incremented by one with each new pass through the equation. In the sequence **[222222]**, **p** equals six, and a maximum of five passes through the equation are necessary, during which **k** would increase from one through five.

In the second clause, the left-hand side of $\leq$ is an accumulator storing vertex sums from the first of the sequence through the vertex in position number **k**. The right-hand side of $\leq$ has two components that are added together to prepare for the comparison between the left and right parts of the clause. The first of these components if completed by multiplying **k** by (**k**-1), and adding that to the result of the accumulator to the right of the "plus" sign. This second accumulator adds the smaller of two numbers; the pointer variable **k**, or the number of vertices in the degree at (**k**+1). The addition takes place in an embedded series of passes through an equation where **i** increases from (**k**+1), and moves step by step to **p**; the end of the sequence.

Consider the sequence **[3331]**. We need to assign concrete numbers to three variables before beginning:

p : The number of members in the sequence
k : Starts at 1; it will be an integer pointer
$\mathbf{d_1}$: The value of the first number in the sequence

The first clause is completed by summing all four members of the sequence: (3+3+3+1=10). Since 10 is an even number, the first clause is satisfied. Then calculate (p-1), 3 in this example; the total number of inequality statements comprising the process in the second clause.

The least confusing way to approach the second clause is to assemble its parts in three places. The first stage is to assemble each value on the left-hand side of $\leq$.

Original Sequence = [3331] $(d_1 = 3, d_2 = 3, d_3 = 3, d_4 = 1)$

Formation of the Second Clause Part One

First pass:	d_1	= 3	=3
Second pass:	$d_1 + d_2$	= 3+3	=6
Third pass:	$d_1 + d_2 + d_3$	= 3 + 3 + 3	=9

Second Clause Part One

First pass:	$3\leq$
Second pass:	$6\leq$
Third pass:	$9\leq$

The right-hand side of $\leq$ has two parts. The first part is formed by the pointer multiplied by itself minus one. The values it adds to the equation are as follows:

Formation of the Second Clause Part Two

First pass:	1(0) = 0
Second pass:	2(1) = 2
Third pass:	3(2) = 6

Second Clause Part Two

First pass:	$3\leq 0+$
Second pass:	$6\leq 2+$
Third pass:	$9\leq 6+$

The final part of the right-hand side of $\leq$ is formed by comparing each pointer value to the number of vertices on the degree to its right, and accumulating the sum.

Formation of the Second Clause Completion

Original sequence = [3331]

		i=k+1	min{k,d_i}	ADD	ACCUMULATOR EQUALS
First pass:	(k=1)	d_2=3	1<3	1	1
		d_3=3	1<3	1	2
		d_4=1	1=1	1	**3**
Second pass:	(k=2)	d_3=3	2<3	2	2
		d_4=1	2>1	1	**3**
Third pass:	(k=3)	d_4=1	3>1	1	**1**

Completion of the Second Clause		
First pass:	$3 \leq 0 + 3$	The inequality system holds up through pass one.
Second pass:	$\mathbf{6 \leq 2 + 3}$	The inequality system breaks down on pass two. Therefore this sequence is not graphable.
Third pass:	$9 \leq 6 + 1$	

The list of graphable sequences in cardinality four after the Erdös and Gallai system of inequalities is:

[3333] [3221] [2222]
[3322] [3111] [2211]

This is the complete listing of all simple degree sequences forming only one connected component, using four pitch-classes.

Degree sequences are an equivalence class not only organizational, but also used to provide information about structural properties of a graph. The most musically suggestive degree sequences, listed in consecutive order, or by the total number of edges required, can be found in the appendices.

Comparisons and Structural Properties

The formalism of graph theory not only helps define structures but also to compare and differentiate them. One of its most powerful and important tools is the concept of isomorphism.

We saw in Figure 1-6 that equivalent graphs can appear different. More specifically, if there is a one-to-one correspondence between the vertices and edge sets of two graphs, then they are called isomorphic. As fundamental as isomorphism is, there is no shorthand way for determining its existence without proving a one-to-one correspondence exists, or for nonisomorphic graphs, that a contradiction in edge sets exists that makes isomorphism impossible.

Consider the three graphs of **[322111]** in Figure 1-11. Figure 1-11A is isomorphic to Figure 1-11B, and the one-to-one correspondence that maps them is listed below. Each vertex from Figure 1-11A with the vertex it corresponds to in Figure 1-11B on the right of each arrow is as follows: <0>→<10>, <1>→<8>, <2>→<7>, <3>→<9>, <4>→<6>, <5>→<11>. This relation can be visualized if one imagines Figure 1-11B rotated so that vertex <10> is in the position of vertex <0>. While all three graphs have the same degree sequence, there is no relation that maps the edge and vertex sets of either 1-11A or 1-11B onto

Figure 1-11C. The contradiction in edge sets that establishes nonisomorphism in this case can be understood by closer examination of the 3-incident vertices <1>, <8>, and <15>.

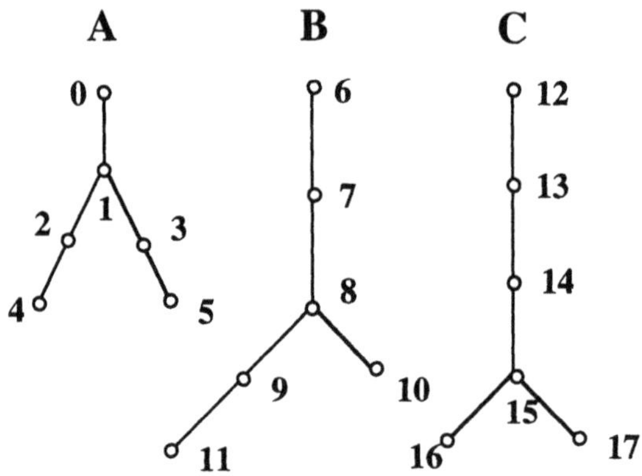

Figure 1-11

Vertices <1> and <8> connect to two different 2-incident vertices, while vertex <15> connects to only one 2-incident vertex. Figures 1-11A and 1-11B are considered nonisomorphic to Figure 1-11C.

One important fact emerges: that a degree sequence does not transmit whether it contains nonisomorphic versions of itself. It is a part of standard working procedure to compare a newly generated graph with any others that share its degree sequence. Isomorphic graphs share structural properties, and although one arrangement may inspire a composer in ways that another does not, there are no significant differences in isomorphic graphs from a structural point of view. Nonisomorphic graphs contain associative differences that affect transformational activity within a graph, and are therefore of extreme value to compositional potential.

Isomorphism has been prominent historically in graph theory, and was an important concept in the development of modern chemistry. There are chemical compounds called isomers, where differing compounds are composed of the same elements. As an example of isomerism, consider the twelve-element substances with the composition C_3H_7OH, or propylic alcohol and Friedel's alcohol, shown below:

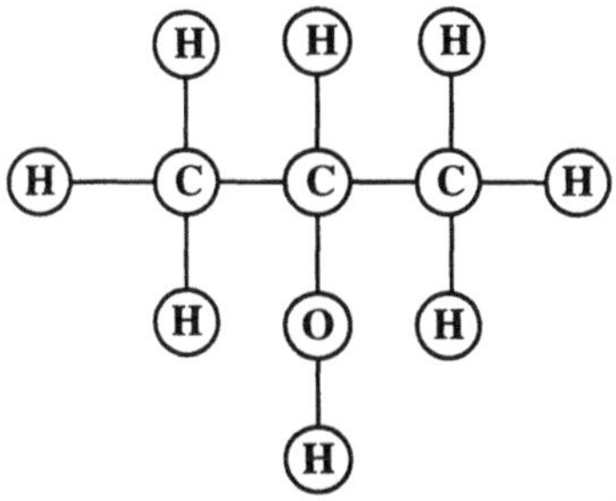

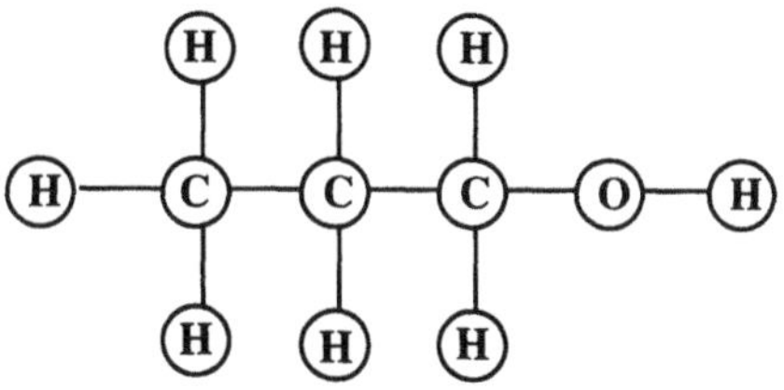

Figure 1-12

Both diagrams in Figure 1-12 are noncyclic, level-three 12-position graphs with degree sequence [444211111111], that could lend themselves easily to musical applications. See page 67.

Saturation

The saturation factor of a transformation graph is the percentage determined from edge sum divided by the edges in the complete graph of its cardinality. It provides a scale of reference by which comparison can be made between graphs of similar or differing cardinalities concerning the completeness of overall connectivity: the smaller the figure the more refined but possibly limited it is. When it is not practical to count edges directly, they can be calculated from a degree sequence. The total number of edges is the sum of all positions in a degree sequence divided by two. The sum of all positions in the degree sequence **[2222]** is eight. The total number of edges is 8/2=4, notated E:4. A graph in which all vertices are connected will have a saturation of one. The saturation factor of Figure 1-2 is calculated as follows: The degree sequence sums forty-eight; there are twenty-four edges. There are sixty-six edges in the complete graph on twelve vertices. Twenty-four divided by sixty-six equals .36; the saturation factor for Figure 1-2 is 36%. Figure 1-8 is 25.7%.

Hamiltonian Cycles

Hamiltonian cycles are named after the famous Irish mathematician William Rowan Hamilton (1805-1865). He marketed a game in the late 1850s that involved finding pathways around a dodecahedron with vertices named after major cities. The object of the game was to pass through each city only once. This type of circuit is one that is musically useful, particularly if the pathway ends on the same vertex on which it began. In this instance the circuit is called a Hamiltonian cycle. If the circuit ends on a different vertex from the one on which it began it is called a Hamiltonian path, and if no such circuit exists, then the graph is called non-Hamiltonian. Much like the presence of isomorphism, the presence of Hamiltonian cycles must be determined or contradicted on a graph by graph basis: very little is known of the underlying nature of Hamiltonian cycles. Figure 1-2 contains Hamiltonian cycles, one of which is <B504A6789321B>

Complementary Graphs

The complement of a transformation graph has the same vertex set as the original. Edges are adjacent in the complement only if not adjacent in the original. The number of levels in a hierarchic degree sequence is always the same as in the complementary degree sequence. Appendix I contains all hexachordal graphs displayed in complementary pairs. In any graph of cardinality six, either the original graph or its complement contains a triangular cycle-class (see Buckley and Harary, 1990: 22).

Matrices

Matrices are nonpictorial representations of structural information contained in abstract transformation graphs essential for computer analysis. Relevant matrices that may yield structural similarities or differences disguised by particular visual presentations include adjacency matrices, incidence matrices, and distance matrices. The matrices described here are adapted from the particularly clear presentation in Buckley and Harary (1990: 117-133).

An adjacency matrix is an array of numbers ZxZ where Z is the vertex set for all nodes and vertices in a graph. Each vertex is compared to every other vertex, listing 1 if adjacent, 0 if not. Row sums indicate the number of objects adjacent from a given vertex, while column sums indicate the number of objects adjacent to a vertex, and are always corresponding numbers. The adjacency matrix for Figure 1-2 reveals an important feature not as obvious in pictorial representation: that all pitch-classes are adjacent to exactly four others.

	0	**1**	**2**	**3**	**4**	**5**	**6**	**7**	**8**	**9**	**A**	**B**		
0	0	1	0	1	0	1	0	0	1	0	0	0	**4**	(Row Sums indicate
1	1	0	0	0	0	1	0	1	0	0	0	1	**4**	Adjacency TO)
2	0	0	0	0	1	0	1	0	1	0	0	1	**4**	
3	1	0	0	0	0	0	1	0	1	1	0	0	**4**	
4	0	0	1	0	0	0	1	1	0	0	1	0	**4**	
5	1	1	0	0	0	0	0	0	0	1	1	0	**4**	
6	0	0	1	1	1	0	0	0	0	1	0	0	**4**	
7	0	1	0	0	1	0	0	0	0	0	1	1	**4**	
8	1	0	1	1	0	0	0	0	0	0	0	1	**4**	
9	0	0	0	1	0	1	1	0	0	1	0	0	**4**	
A	0	0	0	0	1	1	0	1	0	1	0	0	**4**	
B	0	1	1	0	0	0	0	1	1	0	0	0	**4**	
	4	**4**	**4**	**4**	**4**	**4**	**4**	**4**	**4**	**4**	**4**	**4**	(Column Sums indicate Incidence FROM)	

Figure 1-13 Adjacency matrix for Figure 1-2

Incidence matrices are similar to adjacency matrices, and calculated in the same manner, except edges need to be adjacent and incident. Incidence matrices organize directional restrictions and motions within a transformation graph, with incidence TO listed on horizontal lines and incidence FROM listed vertically. Compare the incidence matrix in Figure 1-14 with the adjacency matrix in Figure 1-13.

	0	**1**	**2**	**3**	**4**	**5**	**6**	**7**	**8**	**9**	**A**	**B**	
0	0	0	0	0	0	0	1	1	0	0	0	0	**2** (Row Sums indicate
1	0	0	0	1	0	1	0	0	1	0	1	0	**4** Incidence TO)
2	1	0	0	1	0	0	0	0	0	1	0	0	**3**
3	0	1	1	0	0	1	0	0	0	1	0	0	**4**
4	0	0	0	0	0	1	1	1	1	0	0	0	**4**
5	0	1	0	1	1	0	0	1	0	0	0	0	**4**
6	0	0	0	0	1	0	0	0	1	0	0	0	**2**
7	0	0	1	0	1	1	0	0	0	0	0	0	**3**
8	0	1	0	0	1	0	1	0	0	0	1	0	**4**
9	0	0	0	0	0	0	0	0	0	0	1	0	**1**
A	0	1	0	0	0	0	0	0	1	0	0	1	**3**
B	1	0	0	0	0	0	1	0	0	1	0	0	**3**
	2	**4**	**2**	**3**	**4**	**4**	**4**	**3**	**4**	**3**	**3**	**1**	(Column Sums indicate Incidence FROM)

Figure 1-14 Incidence Matrix for Figure 1-4C Lower Right Diagram

A distance matrix displays the inclusive number of vertices crossing the shortest distance between all vertices. Distance sequences contain a representative of each distance listed in descending order with the total number of occurrences of the distance (above one) listed as a superscript. The distance sequence for each row of Figure 1-15 is: $[3,2^6,1^4]$. When all rows share identical distance sequences, distance patterns for each vertex are permutations of one another. Graphs having this characteristic will be called distance invariant. The diameter of a transformation graph is the highest number in a distance graph; in Figure 1-15 the diameter is three.

	0	**1**	**2**	**3**	**4**	**5**	**6**	**7**	**8**	**9**	**A**	**B**	
0	0	1	2	1	3	1	2	2	1	2	2	2	$[3,2^6,1^4]$
1	1	0	2	2	2	1	3	1	2	2	2	1	$[3,2^6,1^4]$
2	2	2	0	2	1	3	1	2	1	2	2	1	$[3,2^6,1^4]$
3	1	2	2	0	2	2	1	3	1	1	2	2	$[3,2^6,1^4]$
4	3	2	1	2	0	2	1	1	2	2	1	2	$[3,2^6,1^4]$
5	1	1	3	2	2	0	2	2	2	1	1	2	$[3,2^6,1^4]$
6	2	3	1	1	1	2	0	2	2	1	2	2	$[3,2^6,1^4]$
7	2	1	2	3	1	2	2	0	2	2	1	1	$[3,2^6,1^4]$
8	1	2	1	1	2	2	2	2	0	2	3	1	$[3,2^6,1^4]$
9	2	2	2	1	2	1	1	2	2	0	1	3	$[3,2^6,1^4]$
A	2	2	2	2	1	1	2	1	3	1	0	2	$[3,2^6,1^4]$
B	2	1	1	2	2	2	2	1	1	3	2	0	$[3,2^6,1^4]$

Figure 1-15 Distance Matrix for Figure 1-2

	0	**1**	**2**	**3**	**4**	**5**	**6**	**7**	**8**	**9**	**A**	**B**	
0	0	2	2	2	1	2	3	4	4	4	5	5	$[5^2,4^3,3,2^4 1]$
1	2	0	2	2	1	2	3	4	4	4	5	5	$[5^2,4^3,3,2^4 1]$
2	2	2	0	2	1	2	3	4	4	4	5	5	$[5^2,4^3,3,2^4 1]$
3	2	2	2	0	1	2	3	4	4	4	5	5	$[5^2,4^3,3,2^4 1]$
4	1	1	1	1	0	1	2	3	3	3	4	4	$[4^2,3^3,2,1^5]$
5	2	2	2	2	1	0	1	2	2	2	3	3	$[3^2,2^7,1^2]$
6	3	3	3	3	2	1	0	1	1	1	2	2	$[3^4,2^3,1^4]$
7	4	4	4	4	3	2	1	0	2	2	3	3	$[4^4,3^3,2^3,1]$
8	4	4	4	4	3	2	1	2	0	2	3	3	$[4^4,3^3,2^3,1]$
9	4	4	4	4	3	2	1	2	2	0	1	1	$[4^4,3,2^3,1^3]$
A	5	5	5	5	4	3	2	3	3	1	0	2	$[5^4,4,3^3,2^2 1]$
B	5	5	5	5	4	3	2	3	3	1	2	0	$[5^4,4,3^3,2^2 1]$

Figure 1-16 Distance Matrix for Figure 3-14

Cycle-Class Analysis

An awareness of cycles contained in a transformation graph is important to the musical development of ideas based on it. All cycles can be listed and summarized in a cycle-class vector. A cycle-class is based on selecting one representative cycle of all those that can be derived from a specific collection of vertices. The importance of this distinction is that in Figure 1-16, the cycles involving the vertex collection <0>, <1>, <2>, and <5> include permutations and retrogrades such as:

<0125>	<1250>	<2501>
<5012>	<0521>	<1052>
<2105>	<5210>	

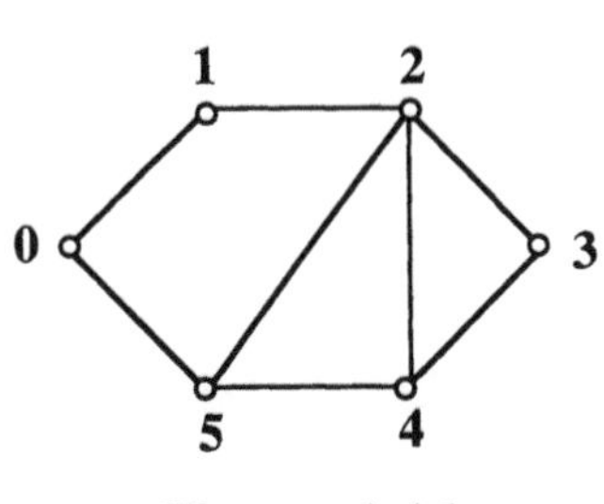

Figure 1-16

A cycle-class is one representative from all possible cycles generated between properly connected vertices. Any of the cycles in the list above could be selected as representing the cycle-class of the vertices <0125>.

A cycle-class vector is a sequence of numbers where each position from one through the cardinality represents the number of cycle-classes of that length in the transformation graph. The cycle-class vector for Figure 1-16 is CV[682211]. Interpretation of each position in this matrix begins with its first position; six cycle-classes involving one vertex. The monocyclic listing is equivalent to the cardinality. The second position indicates eight reflexive cycle-classes involving two vertices; the number of edges. The third position indicates two cycle-classes involving three vertices. These trichordal cycle-classes are <245> and <234>. The remaining positions indicate two cycle-classes on four vertices, <0125> and <2345>, one on five <01245>, and one on six vertices <012345>.

Nonisomorphic graphs can also have identical cycle-class vectors. The hexachords in Figure 1-17 could be realized as a system with complementary hexachords assigned to vertices of both halves of the design.

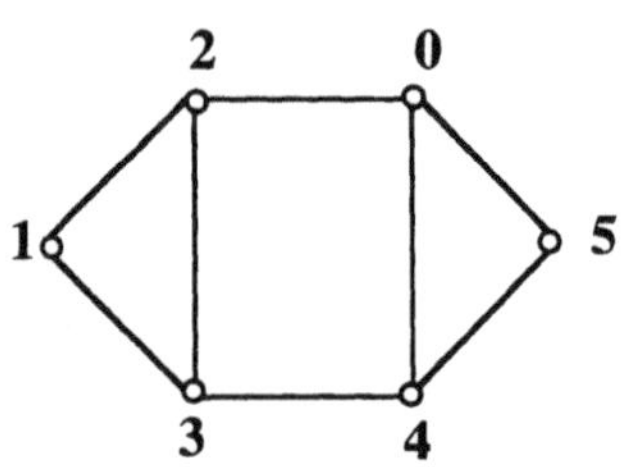

CV[682121] Cycle-Classes:
<0:1:2:3:4:5>
<12:20:05:54:43:32:04:31>
<123:054>
<0234>
<12043:50234>
<120543>

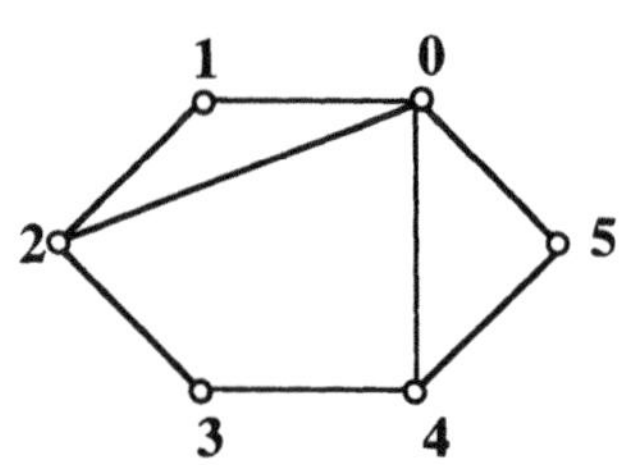

CV[682121] Cycle-Classes:
<0:1:2:3:4:5>
<12:20:05:54:43:32:04:01>
<012:054>
<0234>
<21043:50234>
<210543>

Figure 1-17

Two different classifications of cycle-classes are of particular musical importance. The first are complementary cycle-classes that form two-partition arrays. In Figure 1-2, the cycle-classes <8301B2> and

<695A74> are complementary cycles. The second important classification of cycle-classes are graph-saturating cycles, where cycles engage all vertices in a graph, with duplication allowed. The {6|6|6|6} cycles in Figure 1-2 discussed in the opening pages; <805A42>, <8B7A93>, <6951B2>, and <647103> are examples of graph-saturating cycle-classes.

Part Two

CONSTRUCTION

Basic Techniques

There are several techniques for constructing graphs deliberately, forming patterns and possibilities that are musically suggestive. Constructing pictorial graphs from degree sequences is the most powerful and practical means of developing these compositional spaces, and is accomplished by starting with empty vertices and connecting edges systematically to them.

Construction From Degree Sequence

An algorithmic approach for constructing graphs from a degree sequence requires amplification of one representational aspect of incidence relations. As each position of a degree sequence lists all edges incident to itself, it represents a vertex in the model to be built. The construction becomes more obvious through temporary simplification of interactions, so proceed by eliminating strategic positions in a degree sequence, and tracking edges associated with them, working until the model becomes manageable. Then construct the graph by substituting a vertex for each degree sequence position eliminated in reverse order, and connect edges in reverse using the tally subtracted from the degree sequence as a map.

Each position in a degree sequence transmits the number of edges incident to the vertex that it represents. Degree sequence positions do not indicate from which specific vertices the edges are incident; this needs to be unraveled as the simplification unfolds. The general procedure consists of removing the leftmost position listed in the degree sequence, and making a new list for every distinct way that all edges indicated by this sequence position can be subtracted from the remaining positions. Vertices cannot be incident in more than one way in simple graphs, so we can subtract no more than one edge from each remaining position in a degree sequence.

To draw graphs from the sequence [32221], begin by removing the 3-incident position. The three edges connected to this position can be pulled from the graph in one of two ways. Either one edge is subtracted from each remaining 2-incident position (Figure 2-1A), or the 1-incident position is subtracted with two of the three 2-incident positions (Figure 2-1B).

A	**B**
[32221]	[32221]
111	111
[1111]	[2110]

Figure 2-1

The result of subtraction in Figure 2-1A, [1111], translates into two components graphed in Figure 2-2A. At this stage it is common to have a system; the next vertex added will bond the structure into a single unit. Add the fifth vertex as in Figure 2-2B, then use the map created from columns in the subtraction tally to create the three 2-incident edges in the degree sequence. Or to put it another way, reverse the subtraction by adding the three 1-incident edges above the line to three of the four sequence edges directly below them in the [1111] line. The reconstruction takes place as in Figure 2-2C.

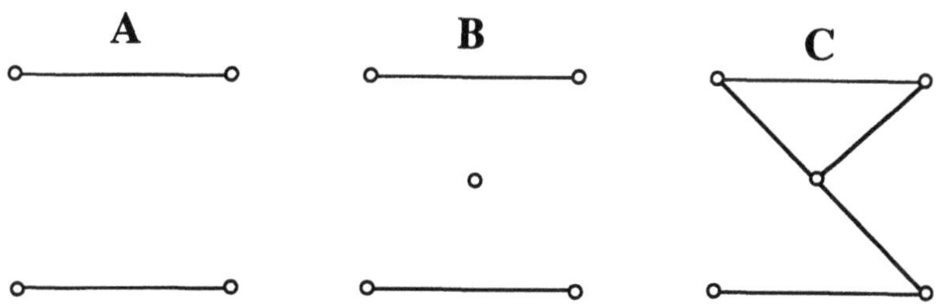

Figure 2-2

To realize Figure 2-1B begin with the two components of Figure 2-3A. Add the fifth vertex as in Figure 2-3B. Add the edges above the [2110] line to specific vertices in each column to the 0-incident vertex and to both 1-incident vertices as in Figure 2-3C.

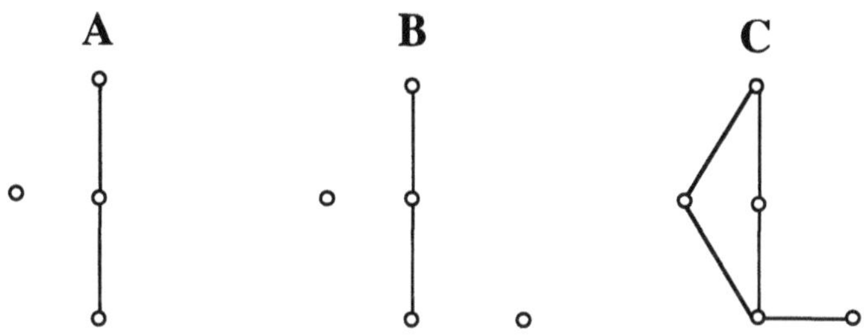

Figure 2-3

The graphs in Figure 2-2C and Figure 2-3C are nonisomorphic. Each has only one cycle-class: in Figure 2-2C of length three, in Figure 2-3C of length four.

If each unique way of subtracting positions from a sequence produced graphs without isomorphism, then a combinational formula could be applied to determine the total number of nonisomorphic transformation graphs that exist for each degree sequence. However, the total number of isomorphisms for each sequence is more elusive, because graphs having distinct subtraction orders that create isomorphisms exist. A method for systematically finding the number of nonisomorphic graphs from a degree sequence has yet to be developed.

Construction From Object Sets

Object set construction differs from degree sequence construction only in lacking an algorithmic approach. Instead, graphs are created based on producing properties and relations of interest. The first technique for object set construction begins with objects having no connecting edges. Figure 2-4 develops a pitch-class network from unconnected objects. It begins the process by listing all objects, in this case pitch-classes from set-class 9-3. A graph that maximizes interval-class one is constructed in Figure 2-4B, and one that maximizes both interval-class one and interval-class three in Figure 2-4C. Only a small portion of the potential intervallic content of the unordered set-class 9-3 is connected. Every pitch-class connects to others either by T_1, T_{11}, T_3, or T_9, creating a focused compositional resource.

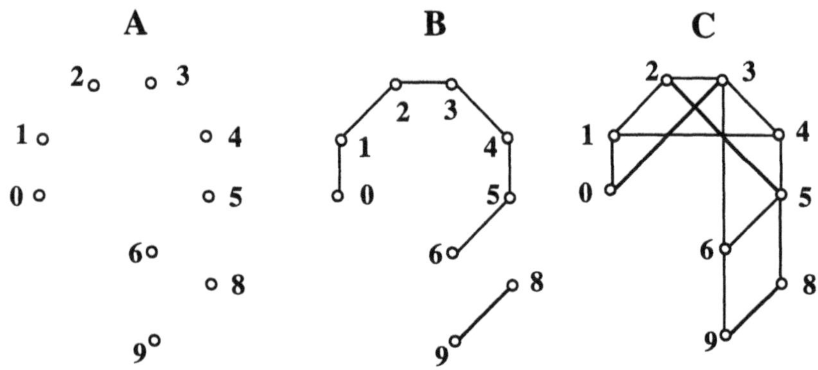

Figure 2-4

The top-down technique begins with every possible connection between objects, and eliminates strategic edges to focus the final product. Figure 2-5A represents all possible connections in pitch-class set 6-7 {012678}, with interval-class vector [420243]. All adjacencies of interval-class one and interval-class four are eliminated in Figure 2-5B. Instead of fifteen interval-class representatives of any unordered hexachord we now have seven.

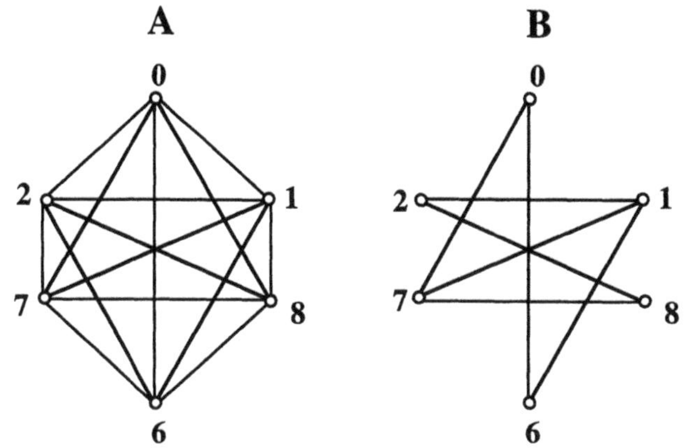

Figure 2-5

Construction of Planar Models

Planar graphs can be created through interaction and superimposition of shapes and patterns. This differs from previous constructions in that one does not produce connections between vertices. Instead, the process of constructing planar graphs starts with edges and produces vertices through intersections; in every location where edges cross, a vertex is produced. Graphs produced in this manner will always be planar.

Planar transformation graphs will be organized into arrangement classes designated: |Order, Generators|:List, with order equal to the number of vertices, and generators equal to the number of cycle-classes that intersect to create graph-saturating cycles. List is the cardinality of each generator listed in nonincreasing order. One arrangement class listing for Figure 1-2 is |12,4|:6666. Equivalence classes can be formed with an equivalent number of either or both members of the ordered pair. The planar generations |6,5| and |8,5| are equivalent generator classes, |7,12| and |7,8| are equivalent order-classes. More than one arrangement class may exist; each one is a key to one method of constructing a particular graph that is of interest. The collection of arrangement classes for a graph offer views that reveal opposing shadows within it.

As an introduction to constructing planar graphs, consider two simple generators both derived from basic figures; the line and circle. The latter is cyclic and bounded, the former linear and unbounded. Graphs can be created through intersection of circles with circles, lines with lines, or circles with lines.

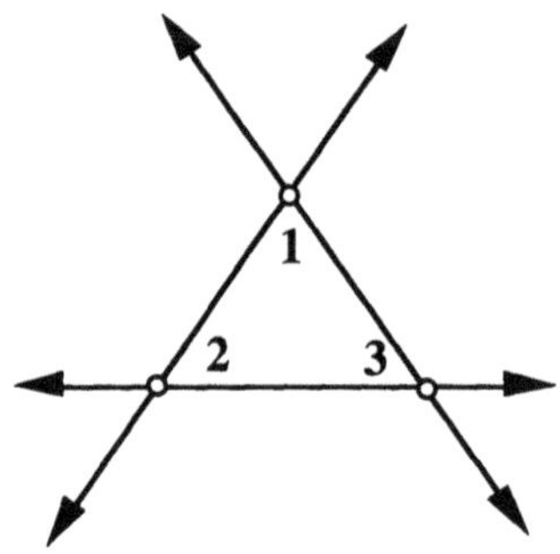

Figure 2-6 I3,3I222

The three labeled vertices in Figure 2-6 are formed from three intersecting lines, partitioning the plane to form a triangular **[222]** graph. Lines that form vertices in this way will be called linear generators; lines are used to generate intersections forming vertices.

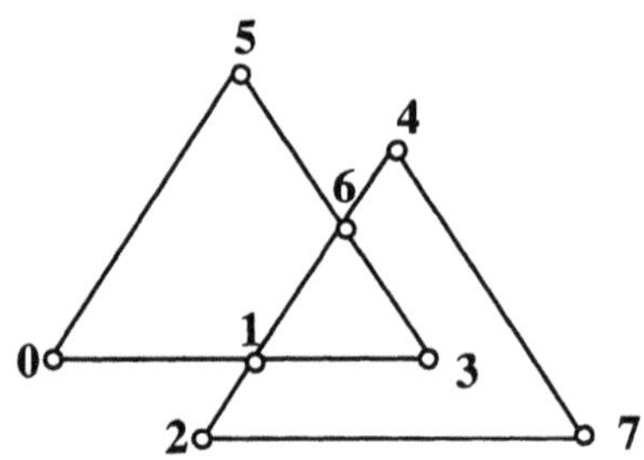

Figure 2-7 I8,2I55

Figure 2-7 can be created through intersection of six linear generators, or developed more directly if lines are grouped into triangles, and the graph is thought of as formed from intersecting triangular generators. Triangular generators are derived from linear generators, as are many other geometric shapes that can be used to create planar graphic models. The overlapping triangular generators in Figure 2-7 produce a smaller resultant triangle. In Figure 2-7, both triangular cycle-classes, <56310> and <47216>, are transpositionally related representatives of set-class 5-13.

A line used as a generator combines with other lines to produce planar vertices within the following extremes:

Let Δ be the number of lines being used to generate vertices
The maximum number of vertices is $1/2(\Delta\ (\Delta-1)$
The minimum number of vertices is $\Delta-1$

Using five linear generators, the maximum number of vertices is ten:

1/2*5(5-1)
1/2 (20)=10

When the maximum number of vertices for five generators is realized, it forms a 3-level cyclic hierarchy **[4443333222]** in Figure 2-8A. The minimum vertices for five generators are four, forming the 2-level hierarchy **[2211]** in Figure 2-8B.

A **B**

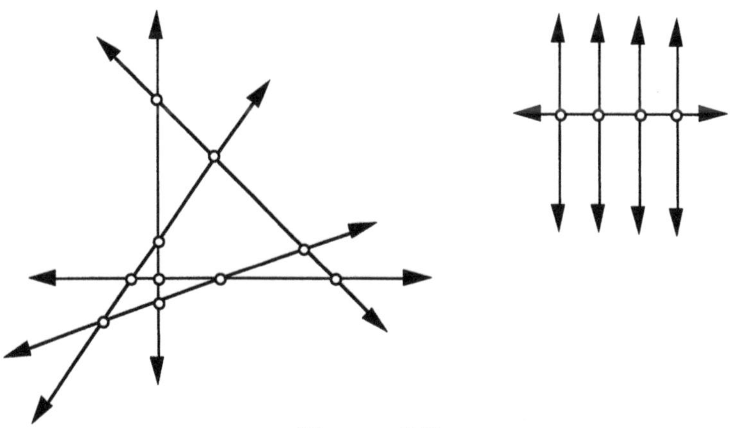

Figure 2-8

A circle used as a generator combines with other circles to produce vertices within the following extremes: the maximum number of vertices for generator X is X^2-X. Maximum number of vertices for generator 1=0, generator 2=2, generator 3=4, generator 4=12, generator 5=20, generator 6=30, generator 7=42. The minimum number of vertices produced using interacting circular generators is one less than the number of generators used.

To generate graphs systematically for a given number of vertices be aware of how many vertices each generator can produce in both extremes, and plan the overall diagram with this in mind. As an example, begin with three generators that are nonintersecting circles. The fourth generator can add anywhere from zero to six vertices. Figure 2-9B uses the maximum intersection to form six vertices.

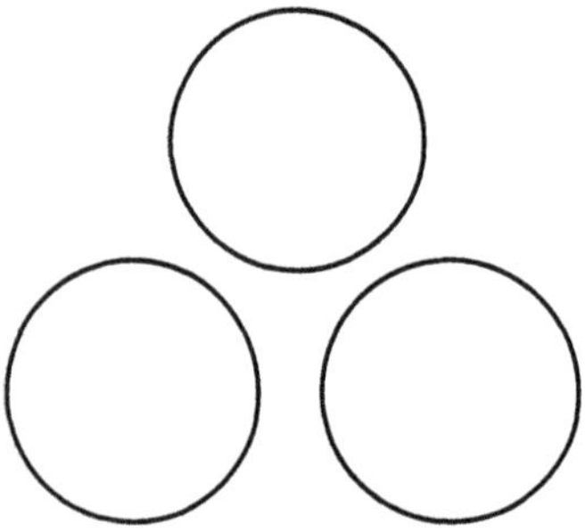

Figure 2-9A

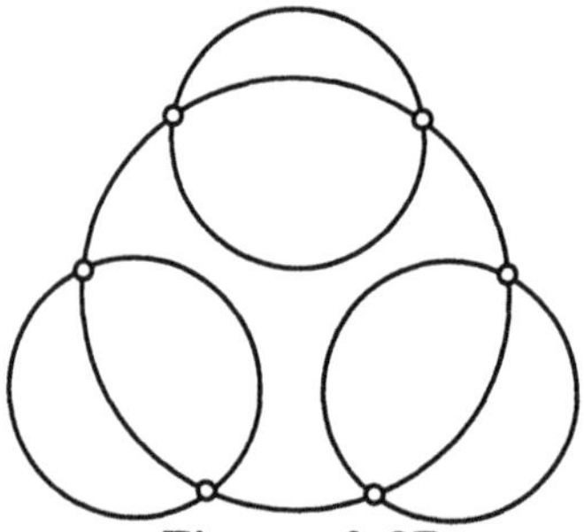

Figure 2-9B

The remaining six vertices can be created several ways, each producing a distinct graph. In Figure 2-9C, the fifth generator could add anywhere from zero to eight zero vertices; adding zero. It is the sixth generator that adds the remaining hexachord by intersecting generator five and two of the three original generators, in Figure 2-9D.

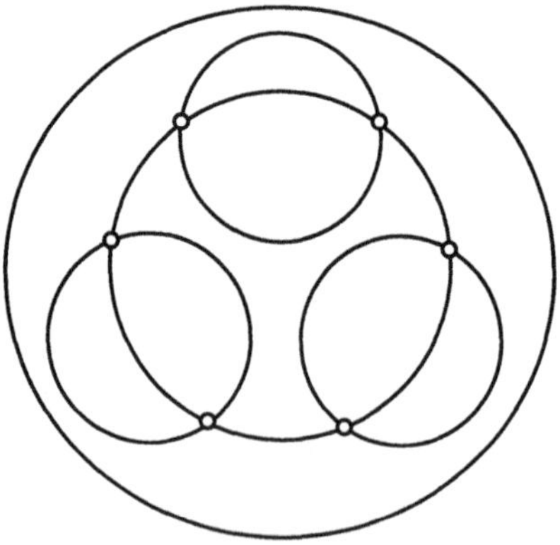

Figure 2-9C

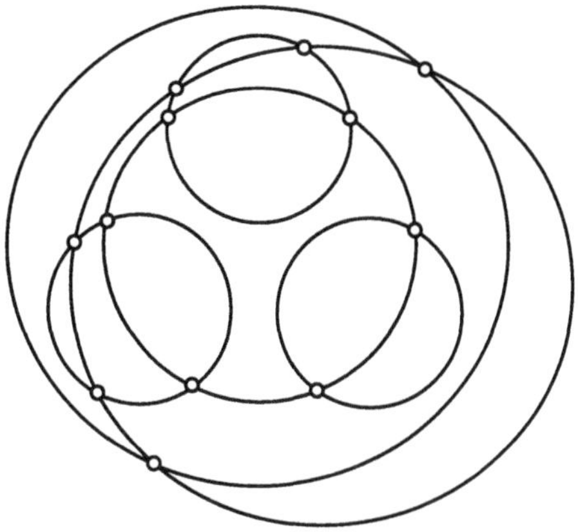

Figure 2-9D

Compare this solution to another with the same three nonintersecting circular generators. Add two vertices with the fourth in Figure 2-10B, four with the fifth in Figure 2-10C, and four with the sixth to produce the planar graph in Figure 2-10D.

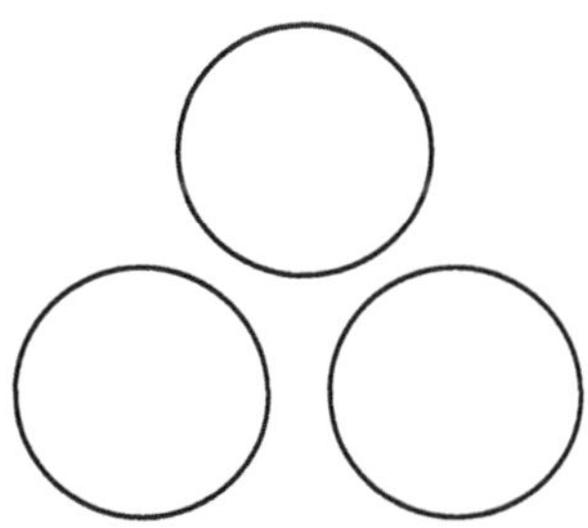

Figure 2-10A

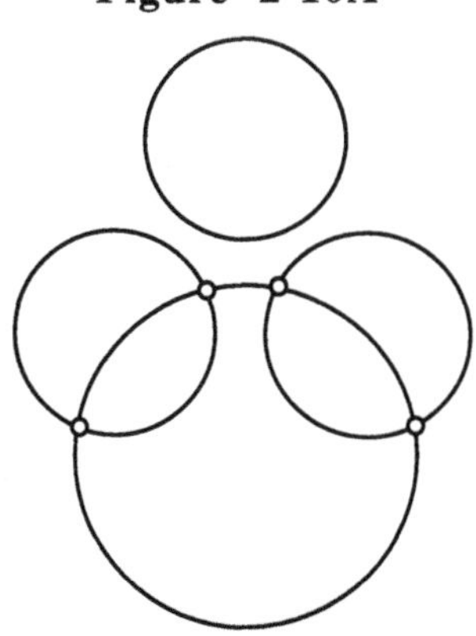

Figure 2-10B

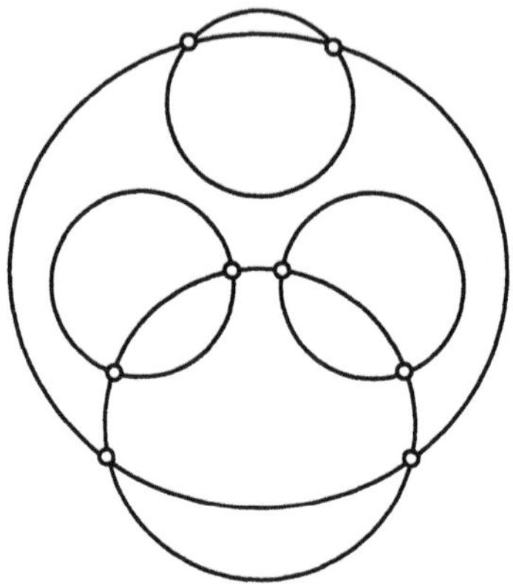

Figure 2-10C

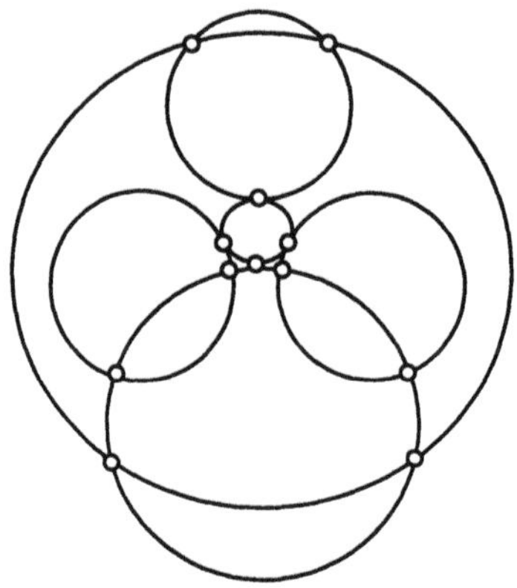

Figure 2-10D

Part Three

CLASSIFICATION

To control graph selection from millions of existing choices, we need to find reasonable and coherent ways to find correspondence between degree sequences, similarities and differences in graphic patterns, and the audible sound of the end-result. Since degree sequence are high-level summaries of all connective possibilities between objects contained in them, classification schemes will be based on information easily obtained from them. When all members of a sequence have access to an equal number of objects, the graph will be called K-regular where K is the number of edges that translate to each of its respective vertices. A hierarchy is a degree sequence where one or many objects have access to an unequal number of objects. A simple observation is required to determine if a degree sequence is K-regular or hierarchic. All degree sequences belong to one or the other, never both, so this classification partitions the list into two parts, both of which also include deeper levels of structural patterning that can now be examined in more detail.

K-Regular Graphs

With uniformly equal numbers in each degree sequence position, K-regular graphs create an abstract symmetry because all objects are equally accessed. In even cardinalities, K-regular degree sequences are possible for all incidence values one less than the cardinality. In odd cardinalities, only even vertex incidence values are possible; odd values create an odd number of odd vertices; and therefore cannot be graphed.

Cardinality	**Possible K-regular sequences**
12	11, 10, 9, 8, 7, 6, 5, 4, 3, 2
11	10, 8, 6, 4, 2
10	9, 8, 7, 6, 5, 4, 3, 2
9	8, 6, 4, 2
8	7, 6, 5, 4, 3, 2
7	6, 4, 2
6	5, 4, 3, 2
5	4, 2
4	3, 2
3	2

The degree sequence **[2222]** is called the 2-regular pitch-class graph on four vertices. Every cardinality greater than two has a 2-regular graph.

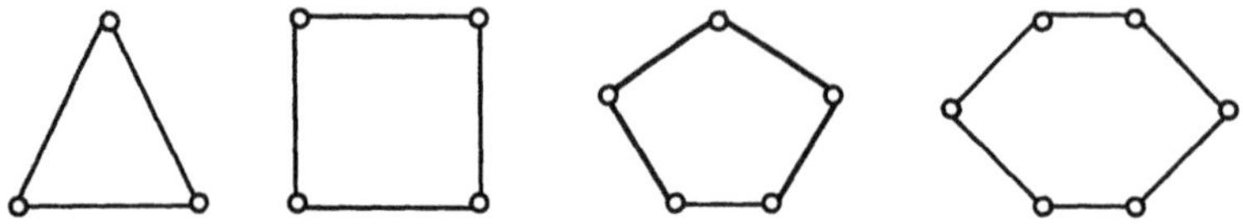

Figure 3-1 Some 2-Regular Graphs

Connective relations of 2-regular pitch-class graphs have the wrap-around structural properties normally associated with rotational arrays. Each forms a single Hamiltonian cycle.

At the other end of the K-regular continuum, when all vertices connect to all others in a pitch-class graph, the properties created correspond to the unordered set-class of the pitch-classes assigned. This type of diagram will be called the complete graph for that cardinality.

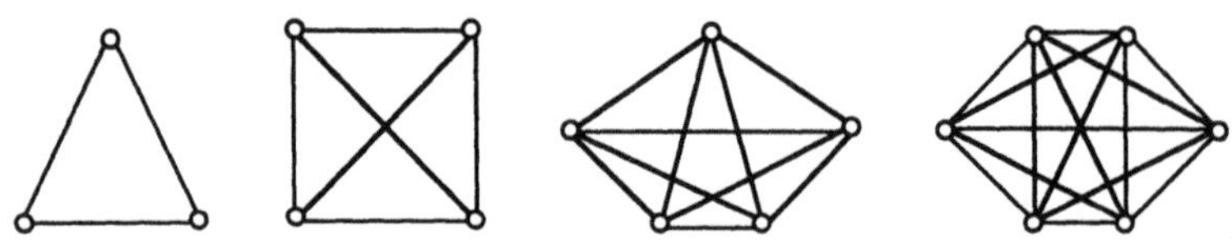

Figure 3-2 Some Complete Graphs

The number of edges for the complete graph in each cardinality is calculated by (K*(K-1))/2 where K is the cardinality.

Cardinality: Edges in Complete Graph

12:66	8:28	4:6
11:55	7:21	3:3
10:45	6:15	2:1
9:36	5:10	

K-4 is the abbreviation for the complete graph on four vertices. K-5 is the complete graph on five vertices, etc. Every other graph of equal cardinality is a subset of the complete graph, having fewer edges.

Many fascinating aggregate distributions involve K-regular formats, perhaps the most inherently balanced of all structures. A description of basic patterns and relationships within some representative examples follows.

[333333333333] 3-Regular Structure

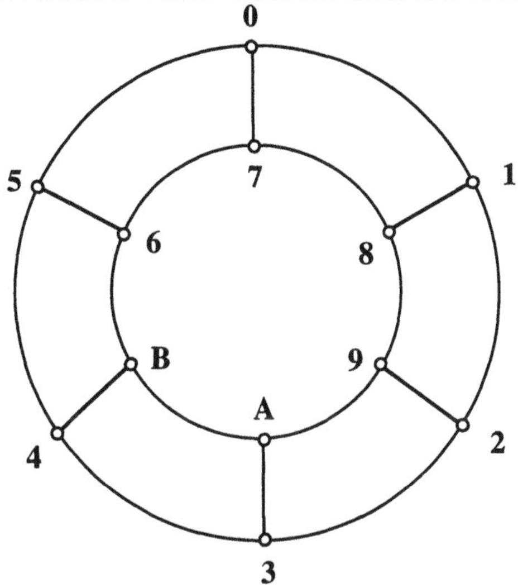

Figure 3-3

A planar 3-regular distribution of the aggregate. The graph is Hamiltonian and has a diameter of four. It contains no odd cycle-classes, and the independent set, <082A46:7193B5>, creates a {6|6} partition. The saturation of this factor-18 graph is 27.2%.

Selected Complementary Cycle-Classes

{2|10}	<07:123456BA98>
{2|10}	<18:2345076BA9>
{2|10}	<29:34501876BA>
{2|10}	<3A:450129876B>
{2|10}	<4B:50123A9876>
{2|10}	<56:01234BA987>
{4|8}	<56B4:0123A987>
{4|8}	<4BA3:50129876>
{4|8}	<3A92:4501876B>
{4|8}	<2981:345076BA>
{4|8}	<1870:23456BA9>
{4|8}	<0567:1234BA98>
{4|4|4}	<0187:23A9:56B4>
{4|4|4}	<1892:34BA:0567>
{6|6}	<012345:6789AB>
{6|6}	<56BA34:298701>
{6|6}	<4BA923:810567>
{6|6}	<3A9812:076B45>

Hamiltonian Cycle-Classes

<078129A34B650>
<01876BA923450>
<0129876BA3450>
<0123A9876B450>
<01234BA987650>
<0123456BA9870>

[333333333333] **3-Regular Structure**

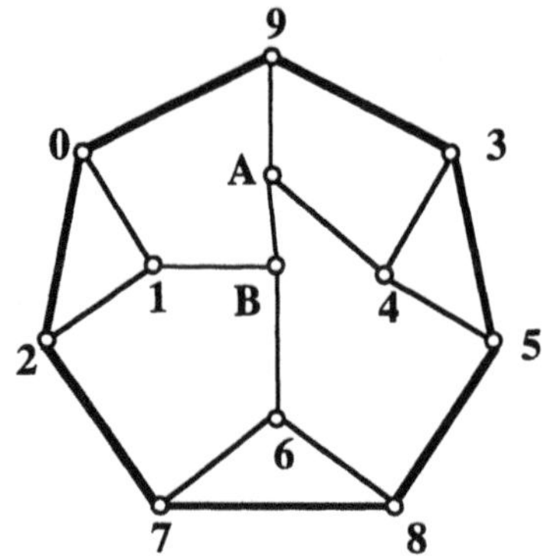

Figure 3-4

A planar 3-regular distribution of the aggregate. The graph is Hamiltonian, <9AB102768543>. The diameter is three. The independent set <09B74:A362:158>, creates a {5|4|3} partition. Several complementary {6|6} partitions in this graph are not cyclic: <345876:21BA90>, <867210:9354AB>, and <86BA45:721093>. The saturation of this factor-18 graph is 27.2%.

Selected Complementary Cycle-Classes

{2|10} <27:01B68534A9>
{2|10} <09:21BA435867>
{2|10} <39:54AB102768>
{2|10} <58:76BA439012>
{2|10} <B6:9A43587210>
{2|10} <A4:935876B120>
{2|10} <B1:9A43586720>

{3|9} <012:9AB678543>
{3|9} <345:872109AB6>
{3|9} <678:539021BA4>

{5|7} <9A453:01B6872>

[333333333333] **3-Regular Structure**

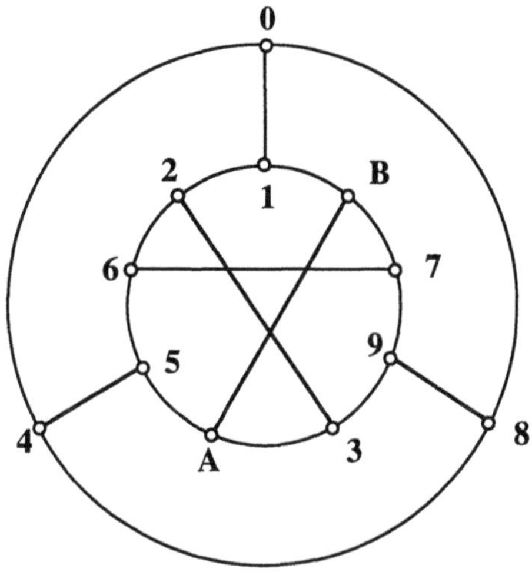

Figure 3-5

A nonplanar 3-regular distribution of the aggregate. The graph is non-Hamiltonian. The diameter is four. The independent set <0B952:461A: 738>, creates a {5|4|3} partition. It contains no {6|6} complementary cyclic partitions, and no tetrachordal cycle-classes. The saturation of this factor-18 graph is 27.2%.

Selected Complementary Cycle-Classes

{2|10} <01:4897BA3265>
{2|10} <45:0893AB7621>
{2|10} <89:401B7623A5>
{2|10} <23:401BA56798>
{2|10} <AB:8456793210>
{2|10} <67:089321BA54>

{3|9} <048:1265A397B>

{5|7} <67B12:045A398>
{5|7} <265A3:04897B1>
{5|7} <B793A:0845621>

{5|7} <26793:801BA54>
{5|7} <21BA3:0456798>
{5|7} <BA567:2398401>

[444444444444] 4-Regular Structure |12,4|:6666

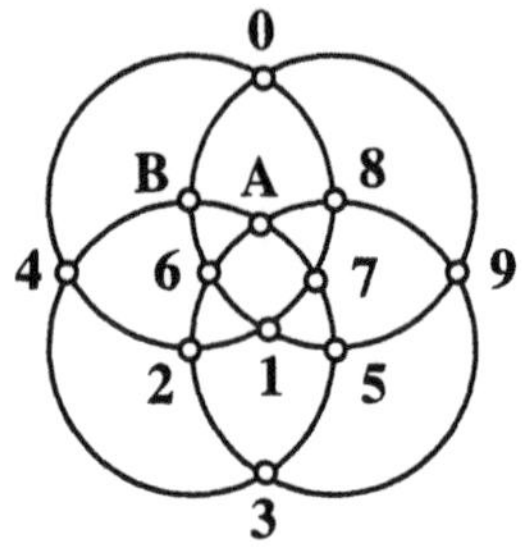

Figure 3-6

Hamiltonian, and the diameter is four. The independent set <0673:4A19: B852>, creates a {4|4|4} partition. The saturation of this factor-24 graph is 36.3%.

Selected Complementary Cycle-Classes

{3|9} <0B4:3216A8759>
{3|9} <089:357AB6124>
{3|9} <935:42178A6B0>
{3|9} <342:0B6157A89>
{3|9} <BA6:043217598>
{3|9} <8A7:904B61235>
{3|9} <517:43908AB62>
{3|9} <261:340BA7895>
{4|8} <0439:BA875126>
{4|8} <A716:0895324B>
{4|8} <0BA8:34261759>
{4|8} <9578:40BA6123>
{4|8} <3215:0987A6B4>
{4|8} <426B:93517A80>
{5|7} <32175:4098A6B>
{5|7} <35162:4BA7890>
{5|7} <4B612:08A7539>
{5|7} <426AB:0871539>
{5|7} <0BA78:9516243>
{5|7} <08A6B:9571243>
{5|7} <98715:326AB04>
{5|7} <957A8:3216B04>
{5|7} <34B62:9517A80>
{5|7} <35789:4216AB0>
{5|7} <4BA80:3261759>
{5|7} <42153:0B6A789>
{5|7} <08759:4BA6123>
{5|7} <0B624:98A7153>
{5|7} <98AB0:3571624>
{5|7} <95123:04B6A78>
{6|6} <04BA89:326175>
{6|6} <908753:4BA612>
{6|6} <342159:0B6A78>
{6|6} <40B623:98A715>
{6|6} <04BA78:326159>
{6|6} <40B612:357A89>
{6|6} <098A6B:421753>
{6|6} <908715:326AB4>
{6|6} <26A871:4B0953>
{6|6} <57AB61:324089>
{6|6} <7AB621:953408>
{6|6} <6A8751:4B0932>

[444444444444]	**4-Regular Structure**	**|12,5|:84444**

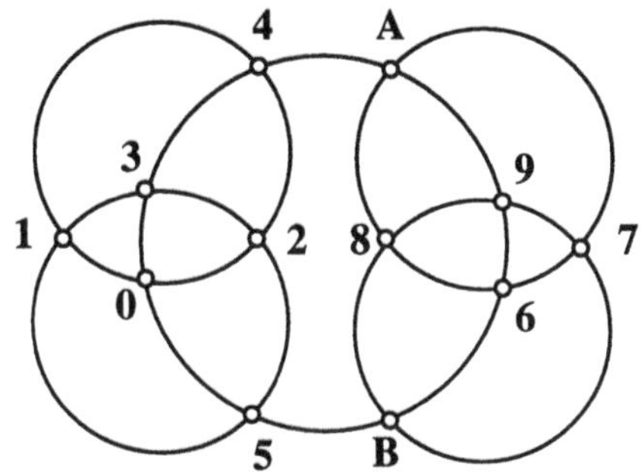

Figure 3-7

This planar, 4-regular distribution of the aggregate is Hamiltonian <431025B8697A94>, and the diameter is three. The saturation of this factor-24 graph is 36.3%. The independent set <1287:40B9:356A>, creates a {4|4|4} partition, and <1287> is literal complement of the cycle-class <4305B69A>. This pitch-class network uses cycles of set class 4-2 related by transposition. The set-class type of intersecting tetrachords and the set-class of the independent set <1287> must always differ. Each dyadic pair would have to exist in three exclusive partitions of the same set-class for this to be possible, and in cardinality four, the third partition always shares a common tone with one of the other two partitions.

Selected Complementary Cycle-Classes

{3|3|3|3} <431:025:A89:B67>
{3|3|3|3} <432:105:A97:68B>
{3|3|6} <310:976:425B8A>
{3|3|6} <320:986:415B7A>
{3|9} <769:134A8B520>
{3|9} <130:79A425B86>
{6|6} <413025:A8B697>

Selected Graph-Saturating Cycles

{4|4|4|4} <4102:5231:B897:A867>
{4|4|8} <1320:7986:4A96B503>

[444444444444]	**4-Regular Structure**

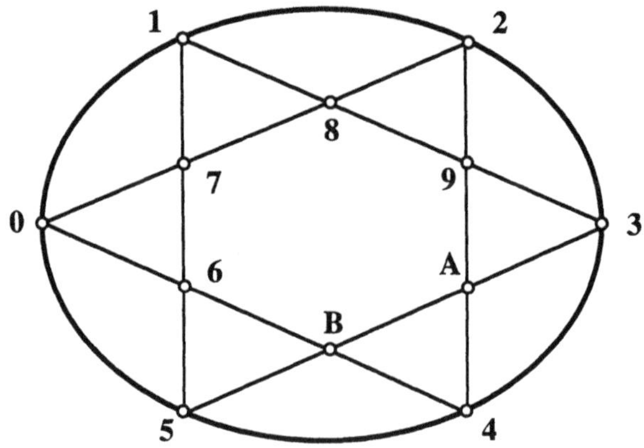

Figure 3-8

The independent set <1649:083B:72A5>, creates a {4|4|4} partition. Planar, and Hamiltonian with diameter of three.

Selected Non-Reflexive Complementary Cycle-Classes

{3|9}	<017:3289AB654>	**{4|8}**	<4AB5:01239876>
{3|9}	<182:05439AB67>	**{4|8}**	<4B65:0123A987>
{3|9}	<293:01876BA45>	**{4|8}**	<5B60:321789A4>
{3|9}	<3A4:0129876B5>	**{4|8}**	<5670:32189AB4>
{3|9}	<4B5:0123A9876>	**{5|7}**	<17650:289AB43>
{3|9}	<560:321789AB4>	**{5|7}**	<18932:076BA45>
{3|9}	<067:32189AB54>	**{5|7}**	<29A43:1876B50>
{3|9}	<178:054329AB6>	**{5|7}**	<3AB54:2987601>
{3|9}	<289:0176BA345>	**{5|7}**	<06B45:3A98712>
{3|9}	<39A:012876B45>	**{5|7}**	<07821:56BA934>
{3|9}	<4AB:012398765>	**{6|6}**	<89AB67:054321>
{3|9}	<5B6:01234A987>	**{6|6}**	<012987:3456BA>
{4|8}	<0671:3289AB54>	**{6|6}**	<123A98:45076B>
{4|8}	<0781:329AB654>	**{6|6}**	<234BA9:501876>
{4|8}	<1782:05439AB6>	**{6|6}**	<012876:345BA9>
{4|8}	<1892:0543AB67>	**{6|6}**	<123987:4506BA>
{4|8}	<2893:0176BA45>	**{6|6}**	<234A98:50176B>
{4|8}	<29A3:01876B45>	**{3|3|3|3}**	<289:3A4:5B6:071>
{4|8}	<39A4:012876B5>	**{3|3|3|3}**	<9A3:45B:067:182>
{4|8}	<3AB4:01298765>		

[555555555555] **5-Regular Structure**

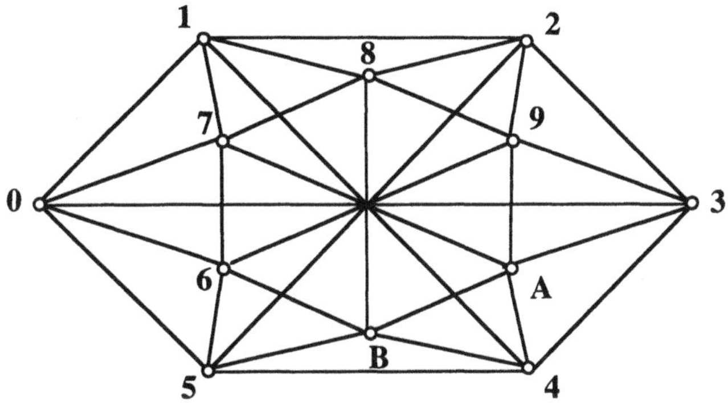

Figure 3-9

Planar and Hamiltonian with diameter of three. The independent set <083B:1694:752A>, creates a {4|4|4} partition. The independent set <083B> has complementary cycle-class <129A45671>. The saturation of this factor-30 graph is 45.4%.

Selected Complementary Hexachordal Cycle-Classes

{6|6} <012987:3456BA>
{6|6} <0123A7:89654B>
{6|6} <012567:8934AB>
{6|6} <0189A7:234B65>
{6|6} <018BA7:692345>
{6|6} <018B67:293A45>
{6|6} <018967:23A4B5>
{6|6} <014B87:23A965>
{6|6} <0143A7:28B569>
{6|6} <014BA7:2396B5>
{6|6} <014567:239AB8>
{6|6} <012876:9345BA>
{6|6} <012396:78B54A>
{6|6} <0125B6:78934A>
{6|6} <014A76:8B5239>
{6|6} <014396:7AB528>
{6|6} <0145B6:3A7829>
{6|6} <0176B5:34A982>
{6|6} <017825:34B69A>
{6|6} <017A45:328B69>
{6|6} <018765:34BA92>
{6|6} <018925:34B67A>
{6|6} <018B45:32967A>
{6|6} <0128B5:34A769>
{6|6} <012345:6789AB>
{6|6} <014B65:3A7829>
{6|6} <014AB5:328769>
{6|6} <012893:7A4B56>
{6|6} <0129A3:4B8765>
{6|6} <017823:9A4B56>
{6|6} <0187A3:296B45>
{6|6} <014A93:2876B5>
{6|6} <078B56:14A932>
{6|6} <07AB56:143982>
{6|6} <07A396:128B54>
{6|6} <078965:123AB4>
{6|6} <07A965:1234B8>
{6|6} <07A345:1296B8>
{6|6} <076925:18BA34>
{6|6} <076B45:189A32>
{6|6} <069325:187AB4>
{6|6} <069AB5:143287>
{6|6} <0698B5:1234A7>
{6|6} <069345:B8217A>
{6|6} <06B825:17A934>
{6|6} <06BA45:123987>

[555555555555] 5-Regular Structure

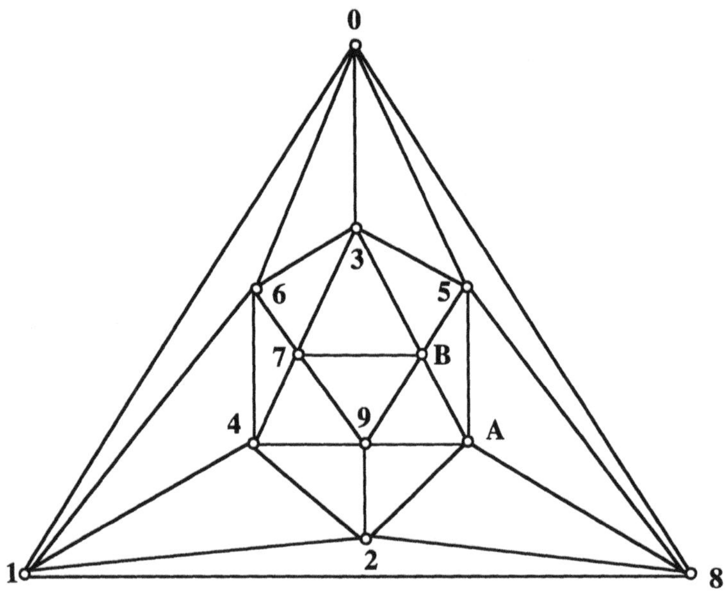

Figure 3-10

A planar 5-regular distribution of the aggregate. The graph is Hamiltonian, and the diameter is three. The independent set <0B2:17A:348:659>, creates a {3|3|3|3} partition. This is a platonic graph: the icosahedron. A platonic graph is planar and regular, where all internal regions are bounded by the same number of edges, in this case three. (Other platonic graphs are the tetrahedron, cube, octahedron, and dodecahedron.) All reflexive cycle-classes have a complementary cycle-class. The saturation of this factor-30 graph is 45.4%. The pitch-class network in Figure 3-10 has at least one trichordal cycle-class from every trichordal set-class.

3-1:<9AB>	**3-5**:<128:016:AB5>	**3-9**: <249>
3-2:<124:467>	**3-6**:<79B>	**3-10**:<036>
3-3:<367>	**3-7**:<035:479:146:58A>	**3-11**:<058>
3-4:<018:29A>	**3-8**:<35B:28A>	**3-12**:<37B>

Selected Complementary Hexachordal Cycle-Classes

{6|2|4} <0149B3:67:5A28>
{6|2|4} <012973:46:5BA8>
{6|2|4} <0129B3:A5:4637>
{6|2|4} <0147B5:36:92AB>
{6|2|4} <0149A5:28:367B>
{6|2|4} <0129B5:A8:3647>
{6|2|4} <0642A5:18:379B>
{6|2|4} <0679A5:3B:1428>
{6|2|4} <0649A8:12:35B7>
{6|2|4} <067928:14:35AB>
{6|2|4} <0379A8:B5:1642>
{6|2|4} <03B928:A5:1476>

{6|3|3} <0149B5:367:A28>
{6|3|3} <0649A5:37B:128>
{6|3|3} <0679A8:142:35B>
{6|3|3} <037928:146:5BA>

{6|6} <014736:829B5A>
{6|6} <014976:82AB35>
{6|6} <012476:8AB735>
{6|6} <018536:2479BA>
{6|6} <018246:379AB5>
{6|6} <0167B3:8249A5>
{6|6} <012463:8A97B5>
{6|6} <0185B3:2A9764>
{6|6} <018A53:29B764>
{6|6} <016735:8AB942>
{6|6} <014635:8297BA>

{6|6} <014285:3679AB>
{6|6} <0128A5:36749B>
{6|6} <018AB5:364297>
{6|6} <014928:367BA5>
{6|6} <0142A8:3679B5>
{6|6} <0679B3:142A58>
{6|6} <067B53:149A28>
{6|6} <064973:12AB58>
{6|6} <0647B3:129A58>
{6|6} <061285:3749AB>
{6|6} <064735:8129BA>
{6|6} <063BA5:147928>
{6|6} <0612A8:3749B5>
{6|6} <067418:35A29B>
{6|6} <0635A8:147B92>
{6|6} <063B58:1479A2>
{6|6} <0379B5:1642A8>
{6|6} <03B9A5:167428>
{6|6} <03BA85:167492>
{6|6} <036418:297B5A>
{6|6} <037B58:1649A2>
{6|6} <037618:249B5A>
{6|6} <035A28:167B94>
{6|6} <05B9A8:163742>
{6|6} <05BA28:163794>
{6|6} <05A928:163B74>

Hierarchies

The vast majority of degree sequences have unequal incidence values, and the method for organizing them is based on this feature. Each different incidence value in a sequence will be counted as a separate hierarchic level. [44332] is a 3-level hierarchy because there are three different values in its sequence; [3322] is 2-level. Hierarchy, in this sense, refers to the number of edges incident to each vertex, not relationships created through pictorial arrangement or musical realization. The maximum extension of this idea using twelve vertices is the 11-level hierarchy with degree sequence [BA9876654321] shown in Figure 3-11; the only 11-level degree sequence with twelve objects.

Classifications of hierarchies are based on the presence and predominance of one-incident edges: acyclic hierarchies, with no cycle-classes, adjunct hierarchies contain cycle-classes and one-incident edges, and cyclic hierarchies containing no one-incident edges.

[BA9876654321]	Level-Eleven Hierarchy

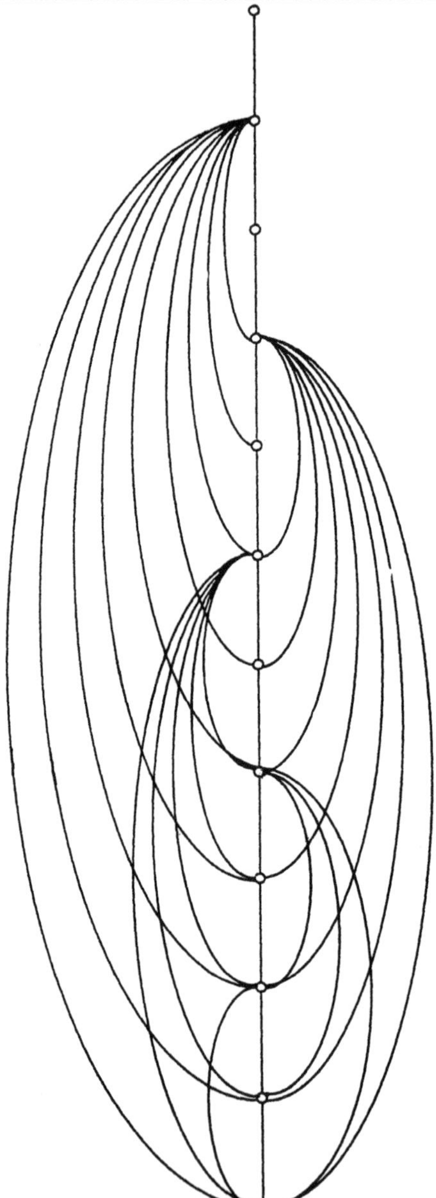

Figure 3-11

Acyclic Hierarchies; Trees

The first classification of hierarchic transformation graphs contains no cycles. Recognizable in having one edge less than the total number of vertices in the graph, they are commonly called trees, but will also be understood as acyclic hierarchies to emphasize their place in the classification scheme of hierarchies. Graph theoreticians focused on trees in the late 19th century after the influential paper published by Cambridge professor Arthur Cayley (1821-1895). There are forty-two distinct acyclic degree sequences for transformation graphs with twelve connected objects, listed on page 159. The number of hierarchic levels and the number of 1-incident edges in an acyclic transformation graph is fundamental to its character. Each cardinality has one tree having exactly two vertices of degree one: the series. The series in any cardinality can be formed from elimination of any single edge of a 2-regular transformation graph. All objects are 2-incident except the first and last. Homeomorphically irreducible trees are those defined by having no vertices of degree two. There are twenty-six homeomorphically irreducible trees on twelve vertices (Riordan, 1958) that can be formed from the twelve acyclic degree sequences having no 2-incident listings using twelve vertices.

[222222222211]	**Level-Two Acyclic Hierarchy**

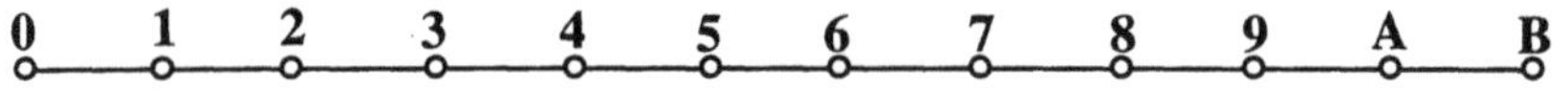

Figure 3-12

The series with diameter twelve, the longest diameter on twelve vertices. The independent set <02468A:13579B>, creates a {6|6} partition. It has a Hamiltonian path, but not a Hamiltonian cycle. The saturation of this factor-11 graph is 16.6%.

[B11111111111]	Level-Two Acyclic Hierarchy

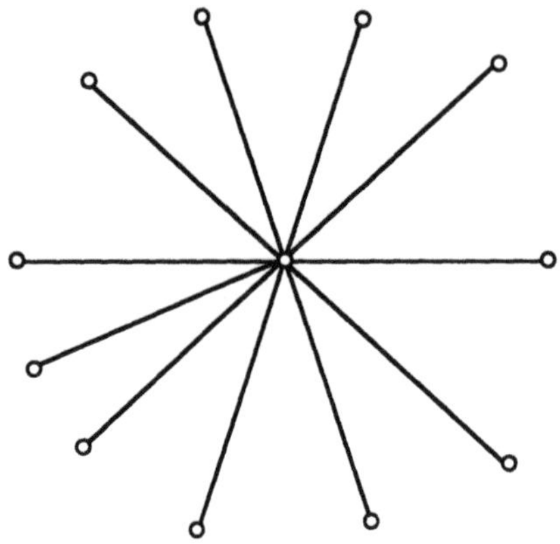

Figure 3-13

One of twenty-six homeomorphically irreducible trees (defined by having no vertices of degree two) on twelve vertices (Riordan, 1958). The diameter is three, and the independent set is the largest possible for a connected graph on twelve vertices: an {11|1} partition. (Adapted from *Breakfast of Champions* by Kurt Vonnegut, page five.)

[543211111111]	Level-Five Acyclic Hierarchy

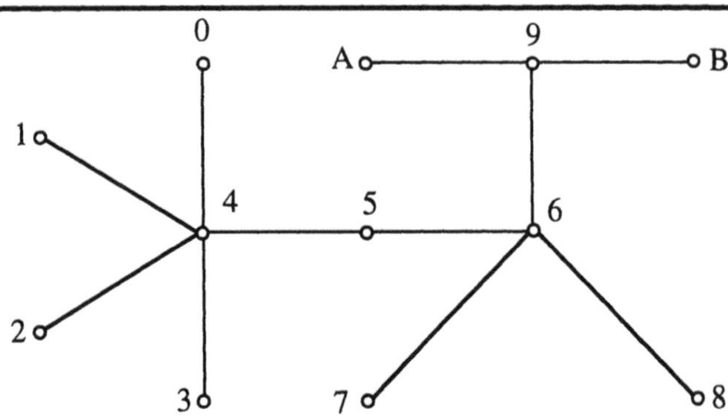

Figure 3-14

The highest number of hierarchic levels among factor-11 degree sequences on twelve vertices, see page 159. The diameter is five.

[552211111111]	**Level-Three Acyclic Hierarchy**

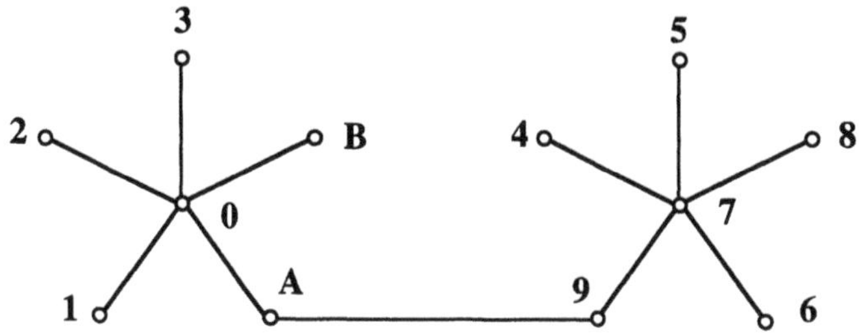

Figure 3-15

The diameter is six. The independent set <094586:7123BA>, creates a {6|6} partition. The saturation of this factor-11 graph is 16.6%.

[444211111111]	**Level-Three Acyclic Hierarchy**

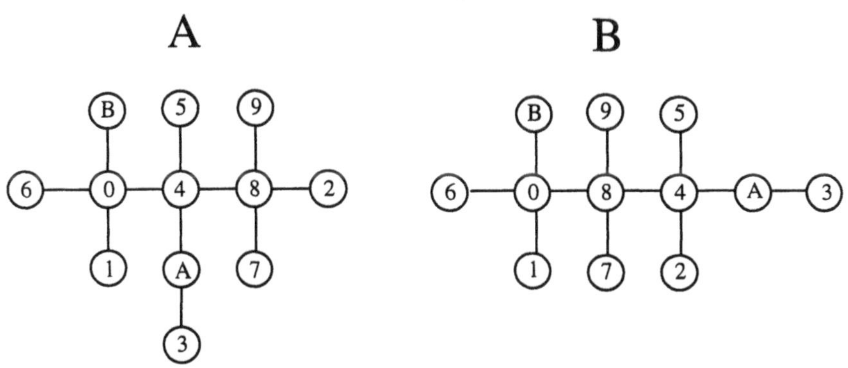

Figure 3-16

These nonisomorphic pitch-class networks are derived from chemical isomers described on page 29. The diameter of network A is five, network B is six. The independent set in network A; <05A8:6B143972>, creates a {4|8} partition. In network B; the independent set <09743: 6B1852A>, creates a {5|7} partition. The saturation of both factor-11 networks is 16.6%.

Adjunct Hierarchies

Adjunct hierarchies can also be found directly from examination of a degree sequence. They are the class of all sequences having at least one 1-incident vertex, in a sequence that is not acyclic. Each of these hierarchies will consist of one primary cycle-class and offshoots or embellishments that can only be accessed from one member of that cycle-class, and the degree sequence will sum greater than twenty-two.

[444411111111]	**Level-Two Adjunct Hierarchy**

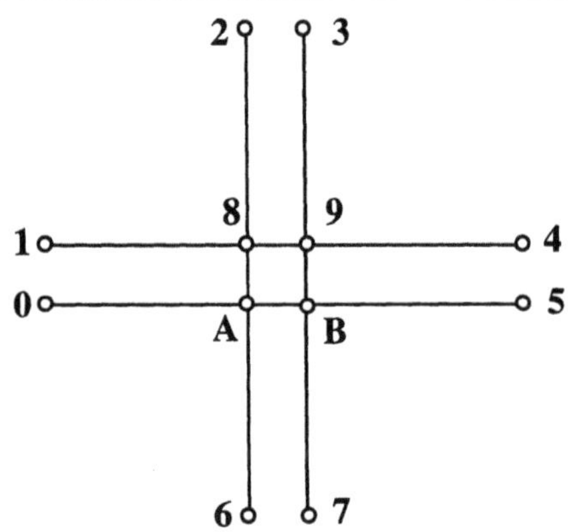

Figure 3-17

An adjunct hierarchy with twelve objects arranged around a central 4-element cycle-class. Each object in this 4-element cycle-class acts as an exclusive gate to two other pitch-classes, and must be sounded in their general context. The diameter is four. The independent set <12975B: 834A06>, creates a {6|6} partition. It is non-Hamiltonian, and the saturation of this factor-12 graph is 18.1%.

[333333111111]	**Level Two Adjunct Hierarchy**

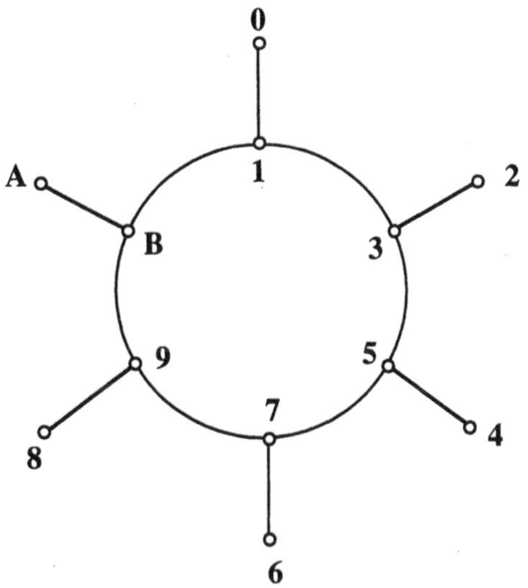

Figure 3-18

An adjunct hierarchy with twelve objects arranged around a central 6-element cycle-class. Each object in this 6-element cycle-class acts as an exclusive gate to a 1-incident pitch-class. The central cycle-class in Figure 3-18 is a whole-tone scale. The diameter is six. The independent set <03478B:12569A>, creates a {6|6} partition. The saturation factor is 18.1%, see also Figure 3-19.

[555111111111]	Level Two Adjunct Hierarchy

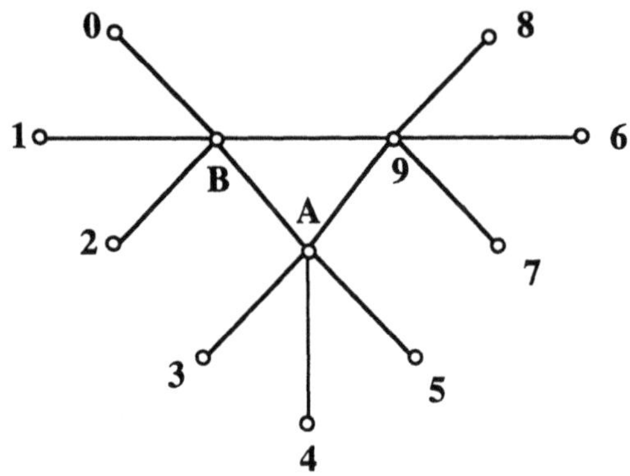

Figure 3-19

An adjunct hierarchy with twelve objects arranged around a central 3-element cycle-class. Each object in this 3-element cycle-class acts as an exclusive gate to three other pitch-classes. The diameter is four. The independent set <012A:9345:B678>, creates a {4|4|4} partition. This factor-12 sequence has a saturation factor of 18.1%, see Figure 3-18.

[AA9333333311]	**Level Four Adjunct Hierarchy**

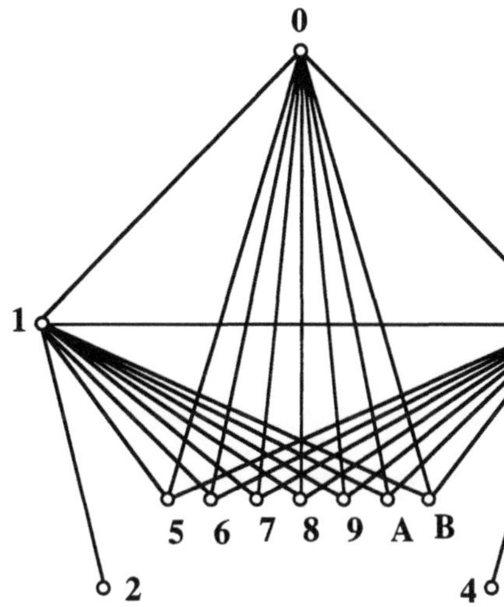

Figure 3-20

The diameter is three. The independent set <0:14:23:56789AB>, creates a {1|2|2|7} partition. The saturation of this factor-26 graph is 39.3%.

Cyclic Hierarchies

Cyclic hierarchies have no 1-incident edges, which is what differentiates them from acyclic and adjunct hierarchies

[444433333333]	**Level-Two Cyclic Hierarchy**

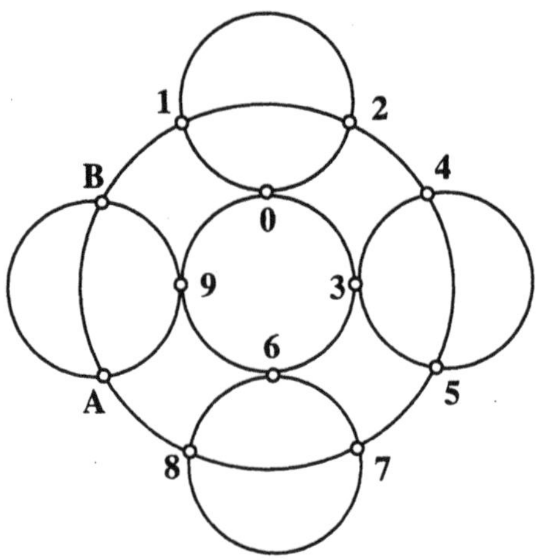

Figure 3-21

The diameter is four, with independent set of size six. The saturation of this factor-20 graph is 30.3%. The independent set <0456AB:123789>, creates a {6|6} partition.

Selected Non-Reflexive Complementary Cycle-Classes

{3|9}	<012:B9634578A>	**{4|8}**	<0369:124578AB>
{3|9}	<345:209678AB1>	**{6|6}**	<124530:B9678A>
{3|9}	<678:A9035421B>	**{6|6}**	<457863:A9021B>
{3|9}	<9AB:103687542>		

Graph Saturating Cycle-Class
{4|4|4|4} <B109:0243:3576:968A>

[444444444422]	**Cyclic Hierarchy**	**|12,5|:66642**

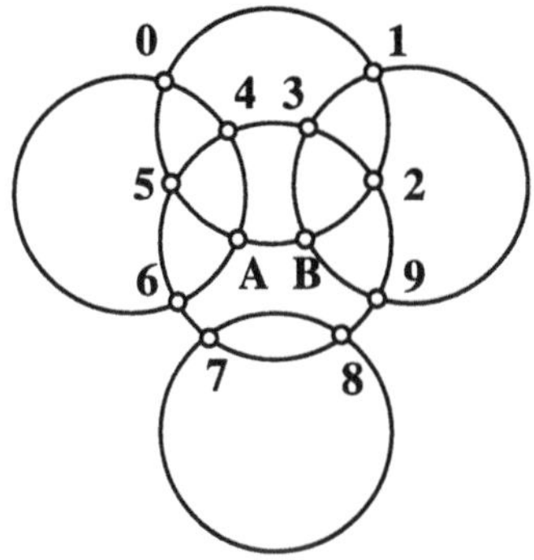

Figure 3-22

This planar transformation graph completes an aggregate using five generators having from two to six vertices. The diameter of this Hamiltonian graph is three. No {4|8} cyclic partitions exist in this graph. The saturation of this factor-22 graph is 33.3%. The independent set <25:0A793: 146B8, creates a {2|5|5} partition.

Selected Complementary Cycle-Classes

{2|10}	<78:0456AB9231>	**{4|8}**	<43AB:01298765>
{3|9}	<045:13BA67892>	**{4|8}**	<0134:5AB29876>
{3|9}	<132:04AB98765>	**{6|6}**	<019876:5432BA>
{3|9}	<45A:06789B321>	**{6|6}**	<6AB987:012345>
{3|9}	<32B:045A67891>		

Graph Saturating Cycle-Classes

{4|4|6} <04A6:13B9:012BA5>

{6|8} <012BA5:43298765>

[443333333333]	Cyclic Hierarchy	|12,5|:77433

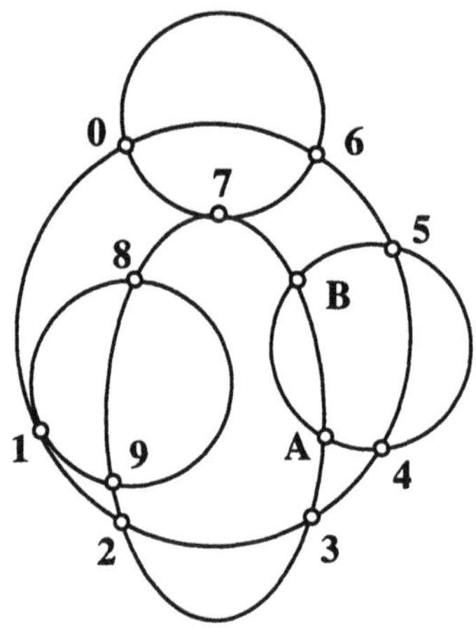

Figure 3-23

This planar transformation graph completes an aggregate using four generators, with three to seven vertices. The diameter is four. The independent set <369:175A:0824B>, creates a {3|4|5} partition. It is Hamiltonian with the cycle <012987BA34560>. There are no reflexive complementary cycle-classes. The diameter is three. The saturation of this factor-19 graph is 28.7%.

Selected Complementary Cycle-Classes
{3|9} <34A:2987B5601>

Graph Saturating Cycle-Classes
{3|3|4|7|7} <076:189:4AB5:2987BA3:0123456>

[443333333333] Cyclic Hierarchy |12,5|:76443

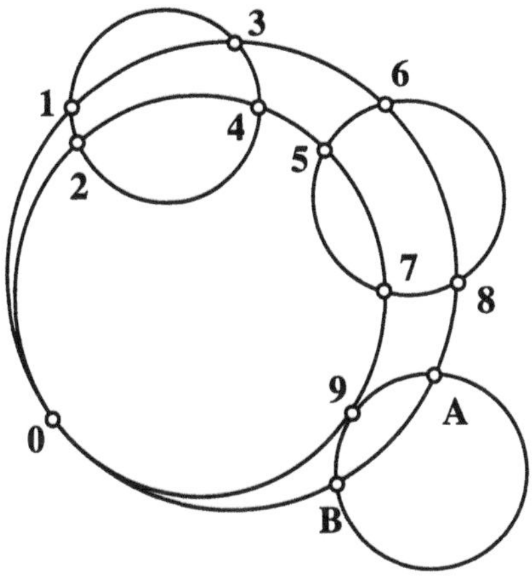

Figure 3-24

This planar transformation graph completes an aggregate using five generators, with three to seven vertices. The diameter is four, with independent set of size five, <23589>. It is Hamiltonian with the cycle <0124365 78A9B0>. There are no complementary cycle-classes in this graph. The diameter is three. The independent set <0A:98532:B1467>, creates a {2|5|5} partition. The saturation of this factor-19 graph is 28.7%.

Graph Saturating Cycle-Class
{3|4|4|6|7} <9BA:1243:6578:024579:01368AB>

[AA2222222222]	**Level Two Cyclic Hierarchy**

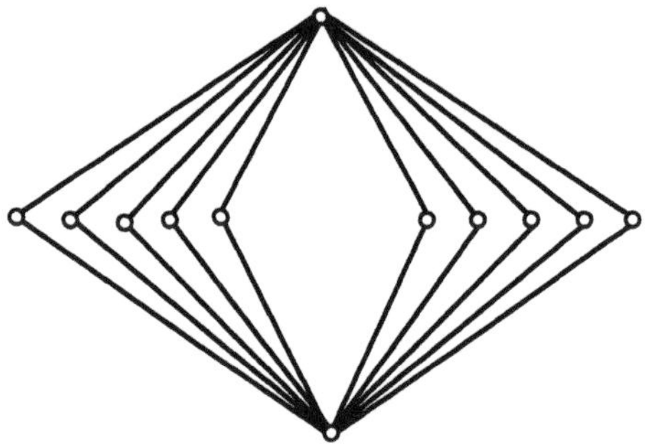

Figure 3-25

This graph contains only cycle-classes of length four. There are forty-five unique tetrachordal cycle-classes. The set-class domain will consist of every tetrachord that contains both of the two 10-incident pitch-classes. The diameter is three. The independent set creates a {2|10} partition. The saturation of this factor-20 graph is 30.3%.

[444444222222]	**Level Two Cyclic Hierarchy**

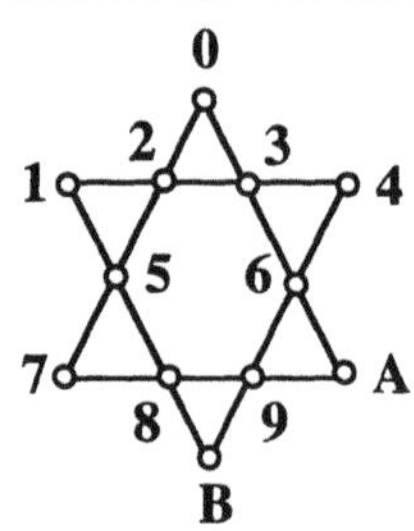

Figure 3-26

The diameter is five. The independent set <056B:13A8:2497>, creates a {4|4|4} partition. The saturation of this factor-18 graph is 27.2%. There are no complementary cycle-classes in this graph.

Graph Saturating Cycle-Classes

{3|3|3|3|3|3} <023:346:6A9:9B8:875:512>

Part Four

REALIZATIONS

Literal Spaces

A preliminary understanding of transformational cycles is necessary for exploration of literal compositional spaces containing properties outlined in this chapter. Transformational cycles are generated by repeating a twelve-tone operator; transposition, inversion, multiplication by five (M), or its inverse (MI), multiplication by seven mod-12, until a loop back to the point of origin occurs. For example, <0>, transposed three half-steps (T_3) produces <3>. From <3>, another transposition by three half-steps produces <6>. From <6>, another T_3 produces <9>, and T_3<9> returns to <0>. This forms a T_3 cycle and is represented counterclockwise, following Morris (1973), in the following way:

		0	
T_3:	3		9
		6	

Two other cycles can be formed using this transformation if the starting pitch is <1> or <2>, and so these transformations are listed next to the first in the complete diagram:

		0			1			2	
T_3:	3		9	4		A	5		B
		6			7			8	

Operations such as T_6, are called reflexive operations, meaning that to repeat the operation is to return to the original entity. T_6<1>=<7>, T_6<7>=<1>. Complex operations are listed using the following notation: T_2I<5>=<9>. It is read "the inversion of F transposed by two half-steps is A." These operations are always understood as taking place mod 12, so that T_7<8>=<3>, and T_7<A>=<2>. M Cycles represent multiplication by five, the cycles of fifths transform. As examples, M<3>=<3>. (Five times 3 is 15, or <3> mod-12 (15-12=3)). MI Cycles represent multiplication by seven, the inversion of the cycle of fifths transform. T_1MI<4>=<5>. (Multiply 4 times 7 = 28; or <4> mod-12 (28-24=4), then <4> transposed one half-step is <5>). For additional information on multiplication as a musical transformation see Howe (1965) and Starr (1978). On the role of multiplication in resolving Z-related sets see Morris (1982b). For application of cycles to musical procedures in general, see Perle (1977a,1977b), Antokoletz (1984), Lewin (1987), and Agmon (1990).

Listing of Reflexive Cycles

T_6:	0	1	2	3	4	5	**T_BI**	0	1	2	3	4	5
	6	7	8	9	A	B	**(Sum:11)**	B	A	9	8	7	6
T_1I	0	2	3	4	5	6	**T_2M:**	0	1	3	4	6	9
(Sum:1)	1	B	A	9	8	7		2	7	5	A	8	B
T_3I	0	1	4	5	6	7	**T_6M:**	0	1	2	3	5	8
(Sum:3)	3	2	B	A	9	8		6	B	4	9	7	A
T_5I	0	1	2	6	7	8	**T_AM:**	0	1	2	4	5	7
(Sum:5)	5	4	3	B	A	9		A	3	8	6	B	9
T_7I	0	1	2	3	8	9	**T_3MI:**	0	1	2	4	6	8
(Sum:7)	7	6	5	4	B	A		3	A	5	7	9	B
T_9I	0	1	2	3	4	A	**T_9MI:**	0	1	2	3	5	7
(Sum:9)	9	8	7	6	5	B		9	4	B	6	8	A

Internal Closure

The compositional space created when complete transformational cycles influence a pitch-class network will be called internal closure. To construct pitch-class networks based on reflexive cycles, arrange interval cycles so that all dyadic pairs contain both members from a reflexive cycle. Dyadic pairs are labeled with equal letters in Figure 4-1.

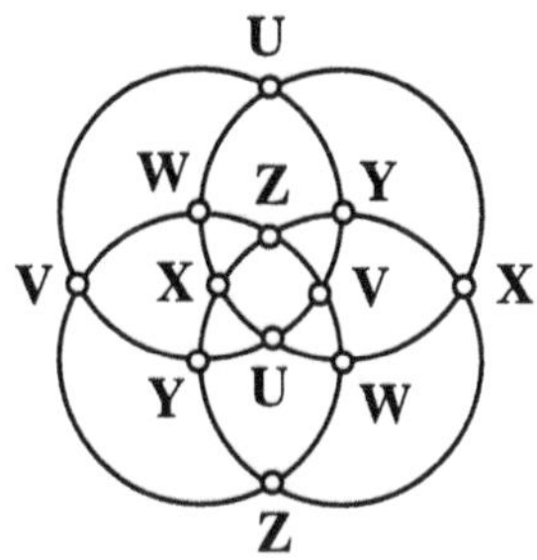

Figure 4-1 Dyadic Pairs Labeling Graph-Saturating Cycle: <UYVUYV: UXWUXW: ZVWZVW: ZYXZYX>

An internal closure based on Figure 4-1 happens when three reflexive transformation cycle-classes are assigned to dyadic pairs within corresponding cycle-classes of the graph-saturating cycle notated in Figure 4-1, forming set-classes within a closed system.

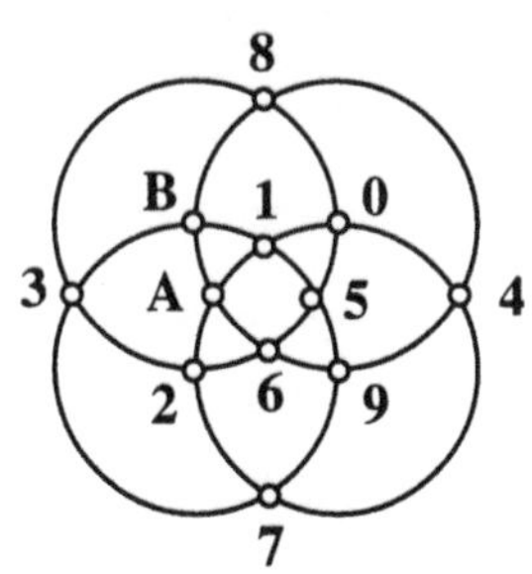

Figure 4-2 Internal Closure on T_2M

The set-classes formed from closure in Figure 4-2; set-class 6-23 <832650>, set-class 6-9 <8496AB>, set-class 6-35 <73B159>, and set-class

6-45 <7401A2>, are the only set-classes that can be formed from internal closure of T_2M. All four set-classes transform into themselves under T_2M.

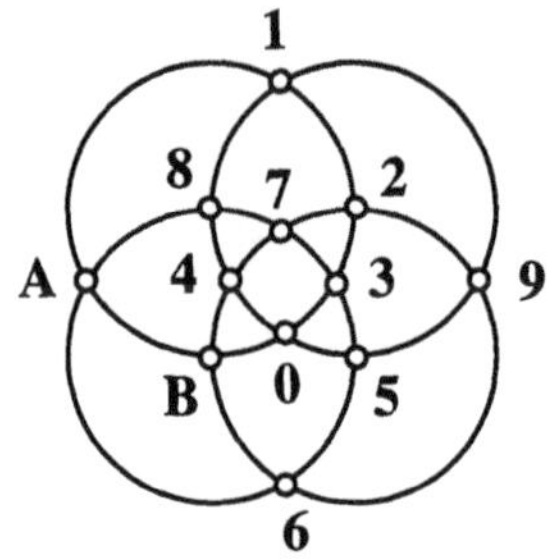

Figure 4-3 Internal Closure on T_1I

The set-classes formed from closure in Figure 4-3; set-class 6-1 <1230BA>, set-class 6-37 <184032>, set-class 6-8 <65378A>, and set-class 6-32 <69274B>, transform into themselves under T_1I.

Calculation of Uniform Generator Cycles

It is also possible to arrange pitch-classes so that all four cycle-classes in the graph-saturating cycle from Figure 4-1 form the same set-class.

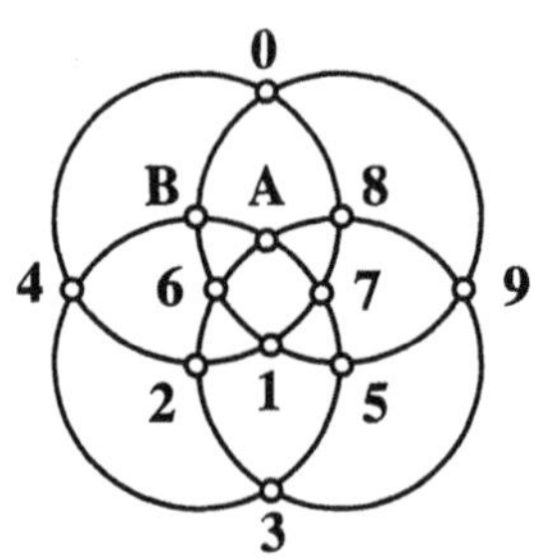

Figure 4-4

All four generators in Figure 4-4 form set-class 6-17: <042178>, <0B6159>, <34BA75>, and <398A62>.

Each generator cycle must share a dyad with each of the other three generator cycles. This abstract musical property will allow partitioning of four all-interval tetrachords within an octachord, paired into two groups of two with no common tones between them. List the complement of the octachord. A partition of this complementary tetrachord into dyads will exist that when added to each of the four partitions of the octachord allows all four resulting hexachords to be of the same set-class. We will see that either dyad partitioning can be added to either of the two groupings of all-interval tetrachords with no common tones. This creates two interrelated diagrams having abstract ZC-related complementary hexachords, with the only difference being the placement of dyads from the complementary tetrachord partitioning. Examine this process in stages.

Choose any all-interval tetrachord. In this example we will choose one from set-class 4-29 and one from set-class 4-15. List every all-interval tetrachord having no common tones on a line below. There will be three for each example.

Stage One

Example One			**Example Two**		
(4-29)	0137		(4-15)	0146	
(4-29)		245A	(4-15)		239B
(4-29)	0137		(4-15)	0146	
(4-29)		2689	(4-15)		3589
(4-29)	0137		(4-15)	0146	
(4-29)		48AB	(4-15)		57AB

The non-intersecting tetrachords built in stage one, considered as a collection, partition an octachord. Stage two partitions this octachord into a second pair of nonintersecting all-interval tetrachords on the next two lines below each example. Each partitioning should use one dyad from both tetrachords in the octachord.

Stage Two

Example One

```
(4-29) 0 1 3 7
(4-29)         2 4 5 A
(4-15) 0   3     4   A
(4-15)   1   7 2   5

(4-29) 0 1 3 7
(4-29)         2 6 8 9
(4-29) 0   3   2   8
(4-29)   1   7   6   9

(4-29) 0 1 3 7
(4-29)         4 8 A B
(4-15) 0   3   4   A
(4-15)   1   7   8   B
```

Example Two

```
(4-15) 0 1 4 6
(4-15)         2 3 9 B
(4-29) 0     6 2     B
(4-29)   1 4     3 9

(4-15) 0 1 4 6
(4-15)         3 5 8 9
(4-29) 0     6   5 8
(4-29)   1 4   3     9

(4-15) 0 1 4 6
(4-15)         5 7 A B
(4-15) 0     6   7 A
(4-15)   1 4   5     B
```

The third step is to find the dyadic partition of the complementary tetrachord that forms hexachords from identical set-classes on each line when each dyad is assigned to one of the nonintersecting all-interval tetrachord collections.

Stage Three

Example One

```
Each Line from Set-Class 6-17
(4-29) 0 1 3 7           6B
(4-29)         2 4 5 A   6B
(4-15) 0   3     4   A   89
(4-15)   1   7 2   5     89

Each Line from Set-Class 6-39
(4-29) 0 1 3 7           AB
(4-29)         2 6 8 9   AB
(4-29) 0   3   2   8     45
(4-29)   1   7   6   9   45

Each Line from Set-Class 6-41
(4-29) 0 1 3 7           29
(4-29)         4 8 A B   29
(4-15) 0   3   4   A     56
(4-15)   1   7   8   B   56
```

Example Two

```
Each Line from Set-Class 6-43
(4-15) 0 1 4 6           5A
(4-15)         2 3 9 B   5A
(4-29) 0     6 2     B   78
(4-29)   1 4     3 9     78

Each Line from Set-Class 6-12
(4-15) 0 1 4 6           27
(4-15)         3 5 8 9   27
(4-29) 0     6   5 8     AB
(4-29)   1 4   3     9   AB

Each Line from Set-Class 6-46
(4-15) 0 1 4 6           29
(4-15)         5 7 A B   29
(4-15) 0     6   7 A     38
(4-15)   1 4   5     B   38
```

An Extension of the Process

Example One

Each Line from Set-Class 6-17			Each Line from Set-Class 6-43		
(4-29)	0137	**6B**	**89**	0137	(4-29)
(4-29)	245A	**6B**	**89**	245A	(4-29)
(4-15)	0 3 4 A	**89**	**6B**	0 3 4 A	(4-15)
(4-15)	1 7 2 5	**89**	**6B**	1 7 2 5	(4-15)

Each Line from Set-Class 6-39			Each Line from Set-Class 6-10		
(4-29)	0137	**AB**	**45**	0137	(4-29)
(4-29)	2689	**AB**	**45**	2689	(4-29)
(4-29)	0 3 2 8	**45**	**AB**	0 3 2 8	(4-29)
(4-29)	1 7 6 9	**45**	**AB**	1 7 6 9	(4-29)

Each Line from Set-Class 6-41			Each Line from Set-Class 6-12		
(4-29)	0137	**29**	**56**	0137	(4-29)
(4-29)	48AB	**29**	**56**	48AB	(4-29)
(4-15)	0 3 4 A	**56**	**29**	0 3 4 A	(4-15)
(4-15)	1 7 8 B	**56**	**29**	1 7 8 B	(4-15)

Example Two

Each Line from Set-Class 6-43			Each Line from Set-Class 6-17		
(4-15)	0146	**5A**	**78**	0146	(4-15)
(4-15)	239B	**5A**	**78**	239B	(4-15)
(4-29)	0 6 2 B	**78**	**5A**	0 6 2 B	(4-29)
(4-29)	14 39	**78**	**5A**	14 39	(4-29)

Each Line from Set-Class 6-12			Each Line from Set-Class 6-41		
(4-15)	0146	**27**	**AB**	0146	(4-15)
(4-15)	3589	**27**	**AB**	3589	(4-15)
(4-29)	0 6 58	**AB**	**27**	0 6 58	(4-29)
(4-29)	14 3 9	**AB**	**27**	14 3 9	(4-29)

Each Line from Set-Class 6-46			Each Line from Set-Class 6-24		
(4-15)	0146	**29**	**38**	0146	(4-15)
(4-15)	57AB	**29**	**38**	57AB	(4-15)
(4-15)	0 6 7A	**38**	**29**	0 6 7A	(4-15)
(4-15)	14 5 B	**38**	**29**	14 5 B	(4-15)

Each of the complementary dyads assigned in stage three is reversible, and creates a ZC-related relationship with the complementary configuration. Unlike Z-relations in other cardinalities, hexachordal Z-relations are also complementary relations, and are therefore called ZC-related to make this feature clear. There are fifteen ZC-related hexachordal set-class pairs. On page 85, configurations come from four intersecting pairs of ZC-related hexachords. The 6-46/6-24 hexachords are formed from a partition of the four all-interval tetrachords within the octachord, where all four tetrachords are from set-class 4-15. The 6-10/6-39 hexachords are formed from a partition of the four all-interval tetrachords within the octachord, where all four tetrachords are from set-class 4-29. Both 6-17/6-43, and the 6-41/6-12 hexachords are formed from a partition of four all-interval tetrachords within the octachord, where one set of tetrachords is from set-class 4-15 and one set is from 4-29. Corresponding configurations from examples one and two transform into one another.

These designs are realized by arranging common dyads onto dyadic pairs of the network:

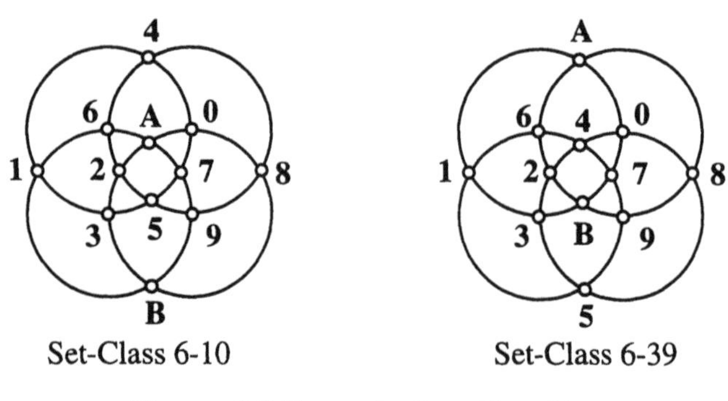

Set-Class 6-10 Set-Class 6-39

Figure 4-5 Example One Case Two

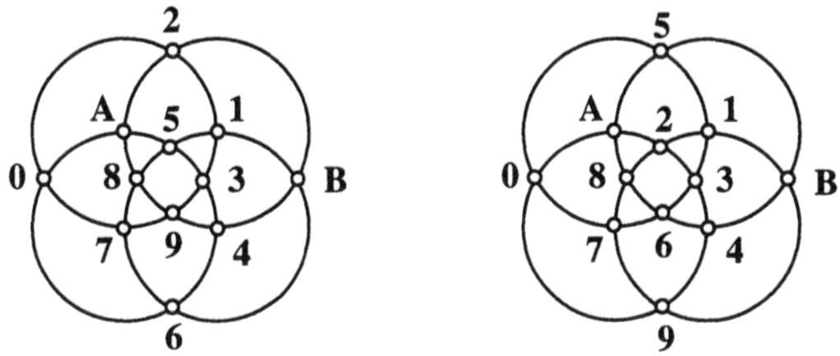

Set-Class 6-12 Set-Class 6-41

Figure 4-6 Example One Case Three

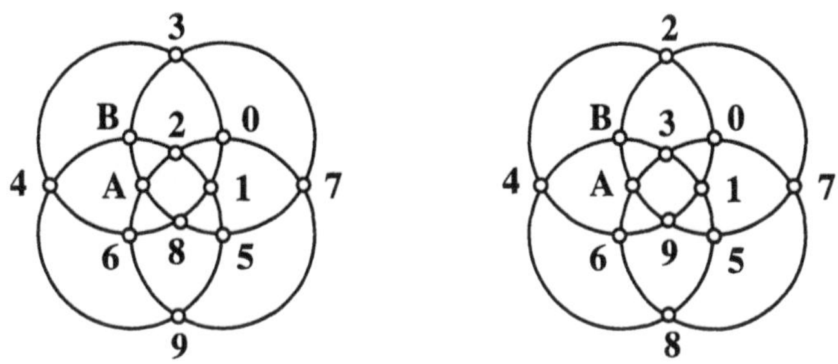

Set-Class 6-24 Set-Class 6-46

Figure 4-7 Example Two Case Three

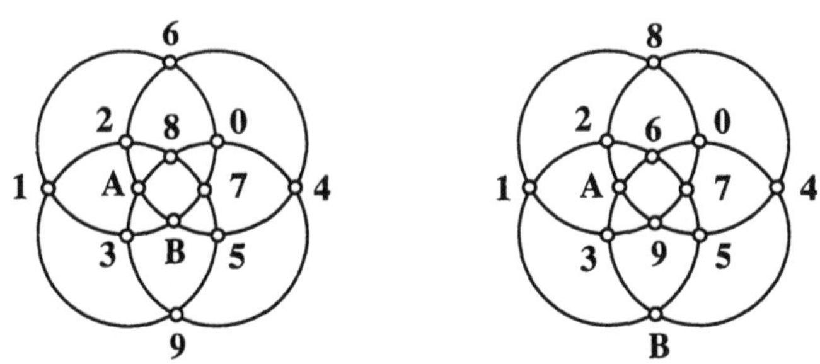

Set-Class 6-17 Set-Class 6-43

Figure 4-8 Example One Case One

Extended Possibilities

Assigning pitch-classes to vertices is one of many possible interpretations of the diagrams come across so far. Many other self-consistent musical constructs could be assigned to vertices in a graph. Vertices could represent set-classes, chordal structures, functional representations, key areas, or unlike, unrelated entities. They could represent written-out musical gestures, rhythms, timepoints, meter, phrase-structure, and could, in general, represent any traditionally serialized musical element such as dynamics, intensities, or articulations. They could represent styles of music, quote musical passages played out like a glass-bead game, or could represent a process or complex of written instructions or indications. They could represent standard transpositional, inversional, or multiplicative transforms linked in complete cycles, or could be connected around various inversional centers. Vertices could represent partitions, and a graph on seventy-seven vertices could be used to determine how all possibilities interrelate. Vertices could represent tempo, or metric rate of speed, extending to phrase-level metric structures.

Structures like pitch-class graphs are single-level graphs, where each vertex contains a single piece of information; a pitch-class. Possibilities also include multi-level graphs, where each vertex contains a complex of information. These include external transformation networks, and the various classes of embedded networks.

Multi-Level Graphs

When vertices of a graph contain embedded graphs, they form multi-level structures. The vertices of these embedded graphs can contain uniform information; such as a single pitch-class. They can also contain diverse materials, either across the graph where each vertex contains a single piece of information, or across the graph where each vertex contains multiple pieces of information.

One class of embedded structures assigns graphs that are related to one another by transposition or inversion of each vertex. When standard twelve-tone operations are applied to a transformation graph, the union of the graph and its transformed shadow creates an external transformation structure. These structures form closure with the cyclic nature of the operators being used. Figure 4-9 is an external transformation structure formed by applying T_2 to the pitch-class network in Figure 4-3. The structure becomes closed as it wraps around the mod-12 universe. The sum of dyadic pairs replicates at the tritone, providing a sub-connection indicated with dotted lines.

Another class of multi-level structures, called embedded structures, are formed from connecting independent components into a single graph, not related by transposition or inversion. Each vertex contains a graph of varying dimension and structural characteristics. An example of an embedded structure can be observed in Figure 4-10.

The final class of structures to be examined contains connected components where each vertex is an embedded copy of the entire structure, extending down into any number of levels: self-embedding structures.

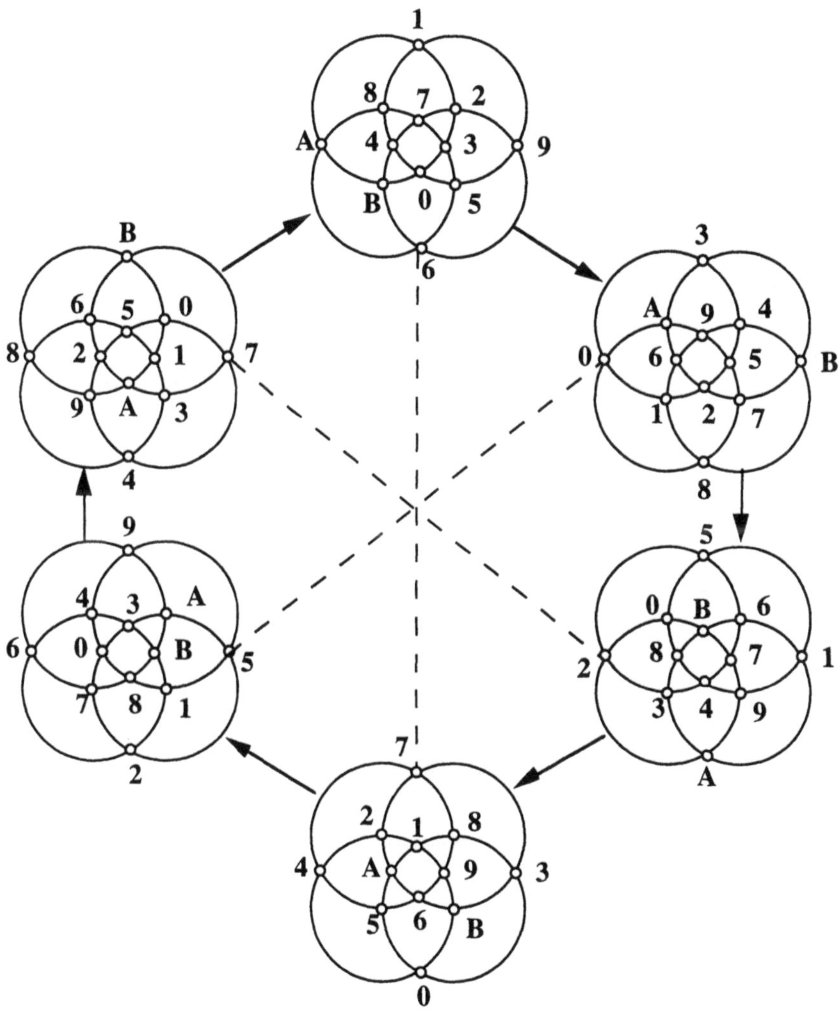

Figure 4-9 An External Transformation Structure

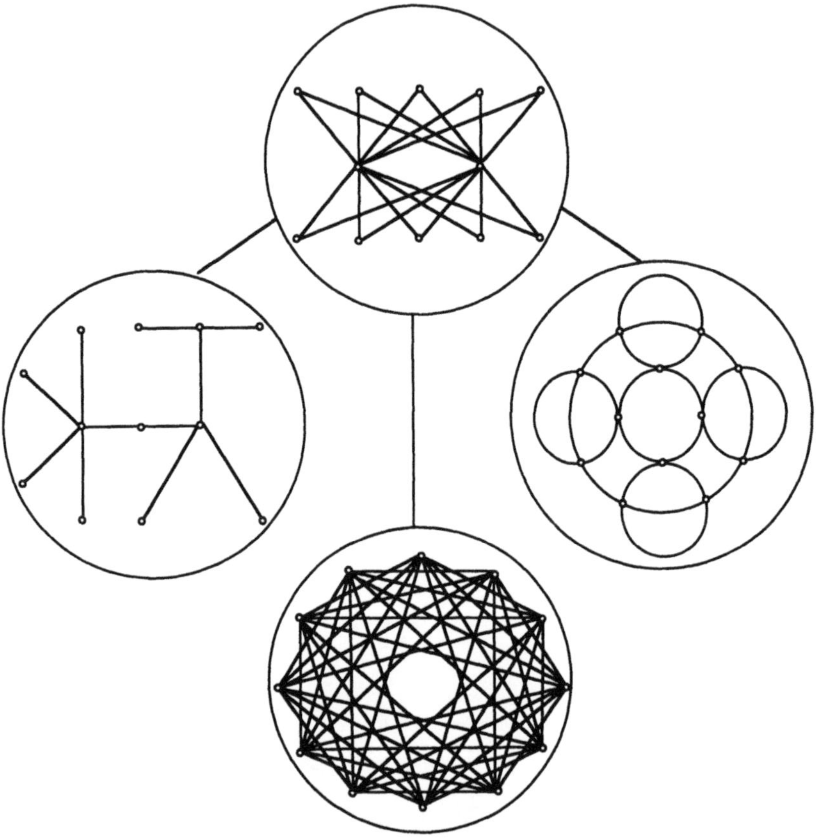

Figure 4-10 An Embedded Structure

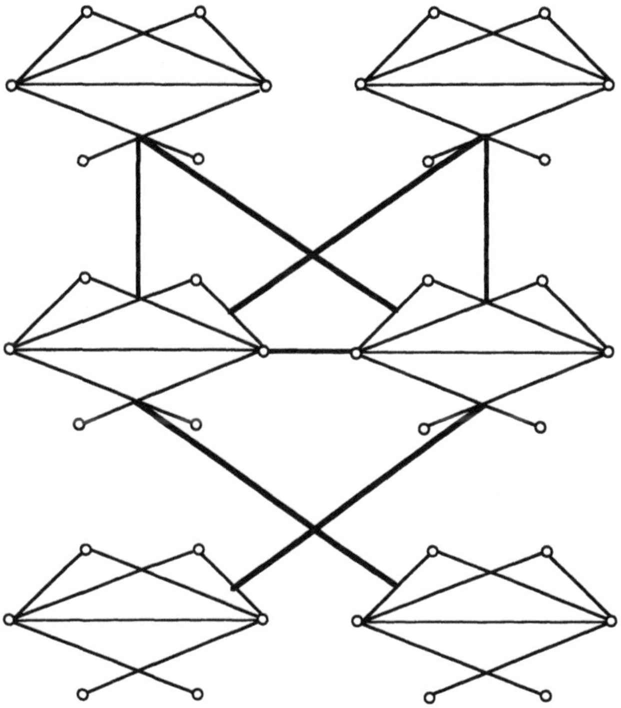

Figure 4-11 A Self-Embedded Structure

“Shock and the Ever-Extending Opposites”; Pitch-Class Graphs and *Broken Sequences*

Pitch-class graphs supplement and refine the ability to articulate orderings within pitch structures in analytical environments. They are of particular value in works by composers such as Stefan Wolpe, whose musical ideas dance in connected fragments, colliding and joining carefully controlled surfaces requiring analytical tools that can cleanly reveal atomic detail.

Form IV: Broken Sequences was written in 1969 at the request of pianist Robert Miller. Wolpe died in 1972, after completing only one additional work. Before examining the opening section, it is important to understand attention given by Wolpe to the mechanism by, and rate at which, the aggregate unfolds. Aggregate completion defines the first sectional boundary in *Form IV*. In the article “Thinking twice,” Wolpe explains that “one may decide to give a certain number of tones a greater preponderance in time, to enhance acutely the plasticity of their pitch arrangements” (p.278). This plasticity comes from controlling association between pitch-classes, using some for cadential patterns, some to create gestural associations, and some for later moments to keep the music sounding fresh and unexpected. He provides two techniques for controlling the relationship of tones: one in which they are understood against remaining tones in the aggregate, and one in which they are not.

> Every pitch constellation smaller than the all-chromatic circuit is either a delay in completing the whole,

> or is an autonomous fragment which can exist outside of the total circuit.

> It may be first unhinged as a part of the total circuit, later hinged back at various degrees of speed, either absorbing stepwise the remaining parts of the circuit, or taking them over in chunks or as a whole. In any way, the modified speed of the all-chromatic circuit points to a concept of modulatory circulation, to an increase or decrease of pitch quantities.
>
> (Wolpe, 1967:287)

The "autonomous fragment" marked "wild" uses set-class 6-12 <024 5AB> exclusively, leaving out all complementary pitch-classes.

In the "pitch constellation" marked "rather slow," pitch-space between the major seventh B and Bb is established in a manner unemphasized on the musical surface. This initially unemphasized musical space becomes a boundary, filled with pitch-classes articulating set-class 7-31 <AB41578> bending in contour and rising in dynamic until the upper limit of the space, Bb, is re-established; an event marked with an accent. Connection among the seven pitch-classes can be graphed using vertices arranged vertically to resonate with their realization.

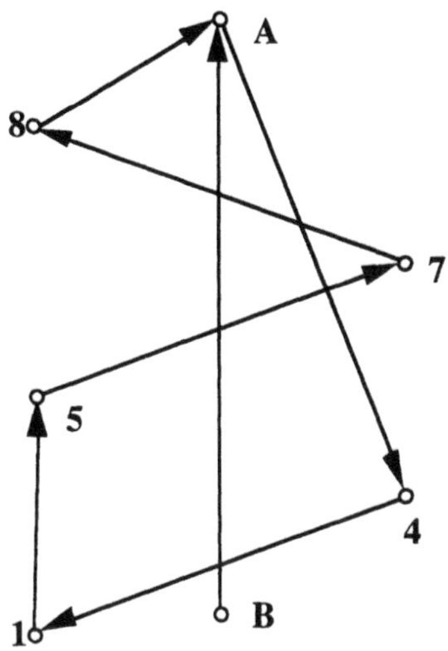

Figure 4-12

Each vertex that fills the boundary seventh in this Hamiltonian path is of degree two, realized in a cycle that excludes the opening <B>. The upper limit of the boundary, <A>, has the highest degree and is given centricity in the realization.

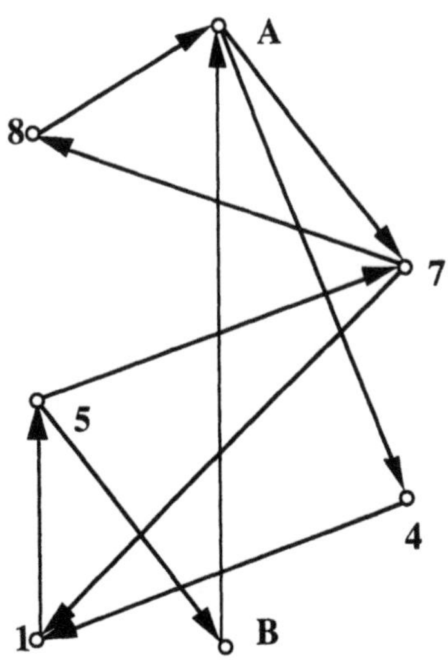

Figure 4-13

The first four barred eighths of the second measure restate a subset of the first measure forming set-class 5-28, <A715B>. They transverse the pitch-space gap from high to low, adding new pitch adjacencies and developing the tritone <A4> from the first measure. The pattern of tritone followed by minor third from the first measure <A41> is inverted <A71>. We will see in later examples that Wolpe commonly maintains a pitch-association in developmental passages like this one. Here, the association <15> is maintained from the previous measure, articulated as a secondary association between the slurred tritones instead of its more pronounced anacrusis position in the first measure.

The pitch-space between the major seventh B and Bb is rearticulated between the fourth and fifth eighths of the second measure. Again, in metrically unemphasized position with the upper limit marked by accent. This final statement of the space is then superseded by presentation of the complementary aggregate pitch-classes <02369>. Pitch-classes <54> break up the sense of literal complement during presentation of these remaining pitch-classes, and articulate the center of the pitch space from <B> to <A>. It is the only material scored in within the originally defined pitch-space and is a previously unarticulated adjacency; simultaneously referential and developmental. Each tone in the complementary aggregate completion is presented higher in pitch than any pitch previously articulated, emphasized both through dynamic and by an increase in upper register. Saving higher registers, and the use of dynamic to articulate arrivals, are also used in *Form IV* to create a sense of line and significance for newly articulated tones of the aggregate.

Dotted lines are used to indicate indirect associations. Indirect associations are based on tangible musical segmentation criteria such as registral similarity, line, or temporal proximity where notation seems to separate, or dissociate them. The dotted <69> association is based on registral association conceived as connecting the line <369>, even though surface notation indicates the line <365>. Dotted lines are also used for temporally adjacent pitches that are notationally nonassociated, as in the case of <54> and <49> both of which contain pitch-classes from distinct lines.

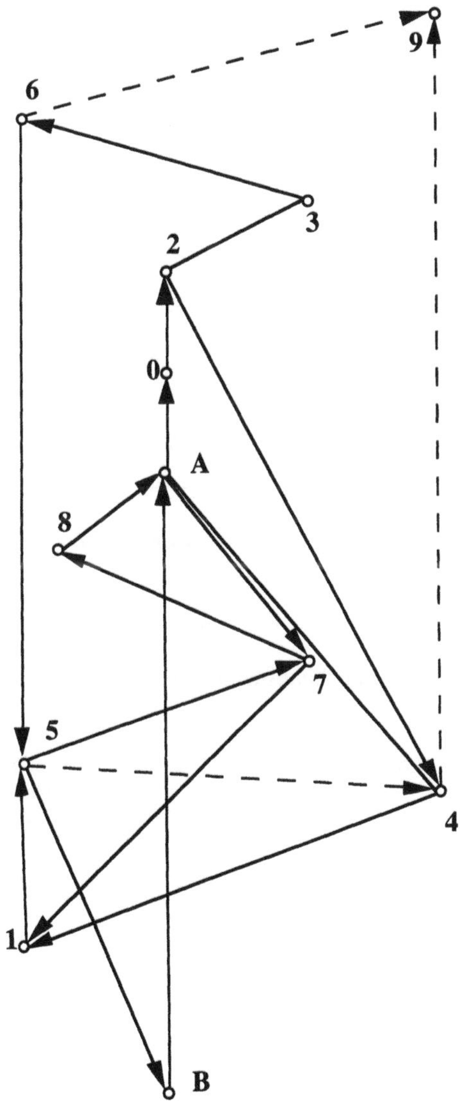

Figure 4-14

Completion of the aggregate is strongly articulated, with simultaneous presentation of the loudest dynamic and highest registral point. It is also isolated as an event from direct connection to its immediate surroundings. The dotted eighth rest supports beaming notation in the other two parts that isolates <9>. Similarities between presentation of adjacencies in this example and in *Form IV* and similarity in registral planning that helps articulate this type of large-scale motion will be apparent.

At the opening of *Form IV*, an eight-position pitch-class space is saturated <AB012345>. Saturation in this sense is the incorporation or sounding of all adjacent half-steps contained within a defined pitch-class space. The relation of this material to the total chromatic is a means of continuation and consequence. The same technique opens the first movement of Wolpe's *String Quartet* (1969), where, after a six-position chromatic pitch-class space is saturated, it takes thirty-one measures before all pitch-classes of the aggregate are sounded. Edward Levy has also described a somewhat similar technique in the second movement of the *Trio* (1964) for flute, cello, and piano (Vinton, 822).

Form IV is a "highly layered" structure enveloped in overlaps between the seven-sectioned tempo scheme and the sectional form of its musical ideas.

Quarter-Note Tempo:	Measures:	Duration in Measures:
112	m1-11	11m
132	m11-43	31m
92	m43-45	3m
112	m46-61	15m
92	m62-65	4m
112	m66-91	25m
92	m91-95	5m

The tempo scheme employs two speeds related to 112 as plus or minus twenty metronome points, with an extended excursion in quicker tempo balanced by three short sections in slower tempo. Superimposed on this tempo scheme is a sectional form defined by gesture, silence, and cross reference. The sectional form interacts with the tempo scheme; at times moving within it, at times against it.

Section **1**:	m1-11	Conflicting forces
Section **2**:	m11-16	Lyric: Contrasting
Section **3**:	m17-30	Developmental: Mechanistic
Section **4**:	m31-45	Transitional: Strong allusions to opening
Section **5**:	m46-61	Emphasis on low register
Section **6**:	m62-78	Lyrical, graceful figures
Section **7**:	m79-91	Increasingly motoric
Section **8**:	m92-95	Triadic: Closing

Cross-referencing within and between these sections is developmental, and ranges from literal repetition to allusion through contour, or microscopic similarities in ordering and adjacency to retrograde and inversion. At various levels from the surface of the music, tracking cross-references and remaining aware of possibilities still unused and fresh leads to deeper structural understanding.

As attention is focused on surface detail comprising aggregate completion in the opening section of *Broken Sequences* consider the ideas, diction, and form of the following quotation from "Thinking Twice."

> a concept of non-gradualness, of non-casual, non-transitional,
> non-process-like thinking leads to a regeneration of expressive
> musical phenomena. It offers the cold, the shabby, the hard,
> the sudden, the inanimate, the rigid, the confused, the joke,
> the excess, the dense, the dropping; the most general, unlayered,
> flat; the extraordinary, highly layered, highly charged; the loose,
> the tattered, the inordinate, nothing much, the perpetual,
> the always interrupted, shock and the ever-extending opposites;
> the simultaneous, the quotational; the noise. (Wolpe, 1967:303)

Somewhat like this quotation, Wolpe's music itself often gathers its electricity from juxtapositions of formal and informal syntax. Also like this quotation, he uses the means at his disposal to create the widest possible variety in presentation.

Figure 4-15 Gesture One and Two, m1-2

The opening gesture establishes connection among eight pitch-classes; set-class 8-1 <AB012345>, divided between the hands:

Right-hand =	<AB0123>	Set-class 6-1	(A Fourth Space)
Left-hand =	<12345>	Set-class 5-1	(A Major Third Space)

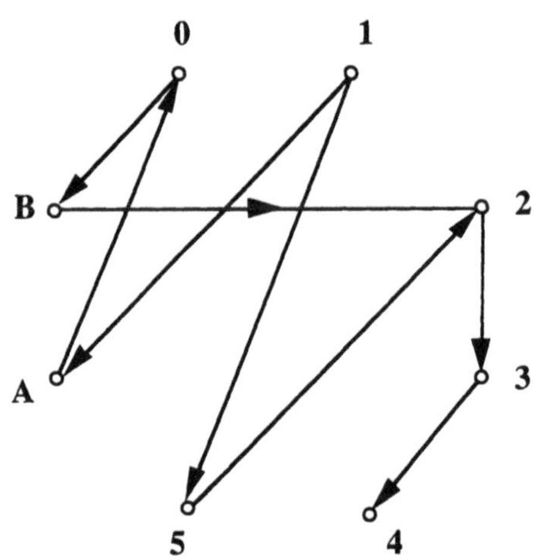

Figure 4-16 Pitch-Class Network of Gesture One

A level-three hierarchy [32222221], the rhythmically independent lines of the right and left hands expose all eight pitch-classes quickly with only one duplicated adjacency <23>. The network is non-Hamiltonian and noncyclic. Non-Hamiltonicity emphasizes the expository, sudden, and fragmented opening. The independent set is size four, partitioning the collection in half, <0124> and <35AB>. Three pitches are displaced by octave, sounding in registers outside the smallest possible pitch space. In the right hand, Bb is displaced above, and Eb is displaced below the pitch-space. This twists the frame of the space, as Bb is the lower boundary and Eb the upper boundary of the opening pitch-class space. In the left-hand, the upper boundary of the major-third pitch-class space, F, is displaced lower. All three displaced pitches belong to a single independent set, and all three are pitch-space boundaries. Displacement of <3> highlights its role as part of the only duplicated association, <23>. Incidence patterns in this example show <2> and <1> to reconcile in the sense that <1> is the only pitch-class 2-incident FROM, and <2> is the only pitch-class 2-incident TO.

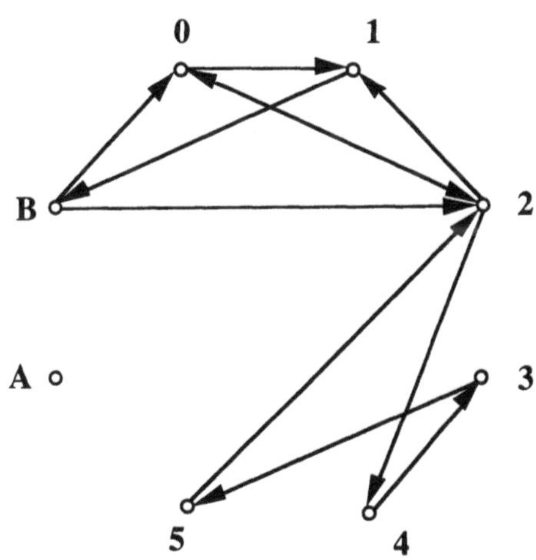

Figure 4-17 Pitch-Class Network of Gesture Two

The second gesture increases the upper-range of the opening by a fourth, and expands in duration from eight to fourteen eighth notes. The first five pitch-classes of the second gesture are an interrupted retrograde of the left-hand of gesture one with modified contour inversion. The ordering <15234> in the left hand of gesture one becomes <43521> in the right hand of gesture two. If the retrograde was not interrupted by the dyad reversal <52> instead of <25>, the <32> and <25> association would have sounded earlier. Perhaps more importantly, rearticulation of the <52> incidence maintains one incidence between gestures, in the same manner as between the hands of the first gesture. Pitch-class <3> is displaced above the space, and pitch-class <0> is displaced below; its presence, a link to the right-hand of gesture one, is formed through transformation of set-class 6-1. Since the first five pitches of the second gesture are based on the left-hand, the opening of gesture two, culminating on the first downbeat of the piece is an embedded reference to both hands of the first gesture. The cadential portion of the second gesture, which commences after the downbeat of the second measure, introduces the first dynamic motion with a crescendo to *mezzo forte*. It forms set-class 6-3 {B02345}, and expands the pitch-class space from a fourth to a tritone <B5>, which is continued in the next gesture.

Incidence in the second gesture is still expansionary. Although several cycle-classes exist in incidence patterns newly created, only <2435> is articulated on the surface of the music. The cycle <435> associates the cadence of the gesture to its opening, where this incidence is first articulated. The degree sequence of gesture two is a four-level hierarchy [53332220] with both higher and lower values in degree sequence positions. Pitch-class <2> is as central to the interplay of connection in the second gesture as it is in the first. Here it is the only 3-incident pitch-class, both 3-incident TO and 3-incident FROM.

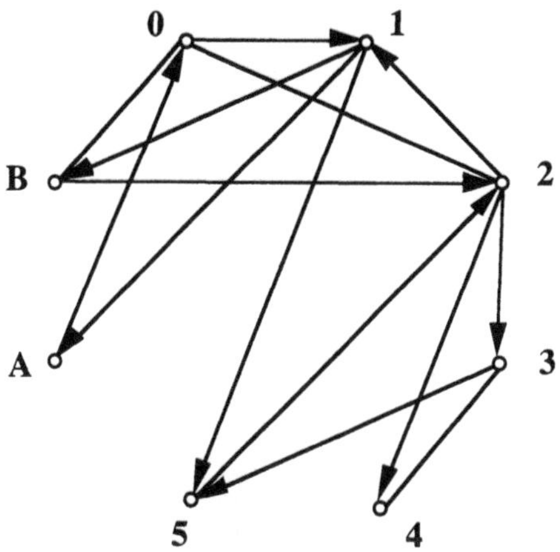

Figure 4-18 Composite Pitch-Class Network of Gestures One and Two

The degree sequence of Figure 4-18 demonstrates degree independence in a 5-level hierarchy [65443322].

	0	**1**	**2**	**3**	**4**	**5**	**A**	**B**	**Incidence TO**
0	0	1	1	0	0	0	0	0	2
1	0	0	0	0	0	1	1	1	3
2	1	1	0	1	1	0	0	0	4
3	0	0	0	0	1	1	0	0	2
4	0	0	0	1	0	0	0	0	1
5	0	0	1	0	0	0	0	0	1
A	1	0	0	0	0	0	0	0	1
B	1	0	1	0	0	0	0	0	2
Incidence FROM	3	2	3	2	2	2	1	1	

The incidence matrix reveals a highly controlled and restricted interaction of pitch-classes on the surface of the music where incidence TO and FROM (except for <4> FROM) is usually less than the degree of the pitch-class. This is further reinforced in that five of the sixteen incidence relations remain 1-incident. With an eight pitch-class environment, an incidence value of three indicates connection to less than half the remaining pitch-classes. Only four of the sixteen incidence relations have values of three or greater, and only one, <2> FROM is greater.

Figure 4-19 Gestures Three, Four, Five, and Six, m3-9

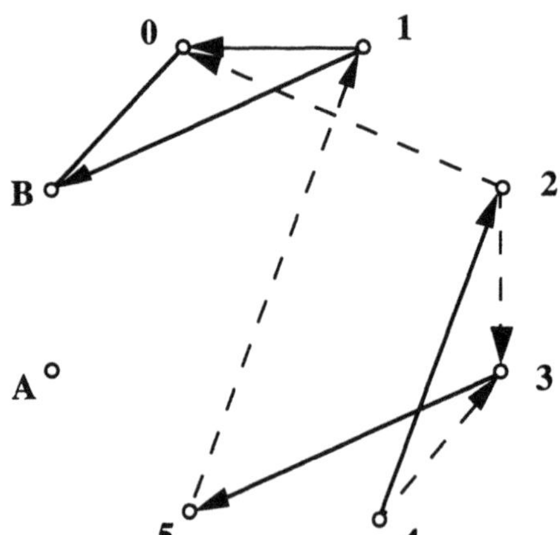

Figure 4-20 Pitch-Class Network: First 7 Eighth-notes of Gesture Three

Gesture three is a variation of gesture two, remaining within the tritone pitch-class space <B5> established there. Expanded in duration to seventeen eighth notes from fourteen in gesture two, it is expanded in register by a minor sixth. It contains the first *forte* of the work, and uses dramatic dynamic motions. There are four displacements, <2> displaced above, <5> displaced below, and the dyad <0B> displaced above. This is the last place in which displacements are embedded within the texture until gesture seven. During the cadence of gesture three, and gestures four, five and six, textures in which there are no displacements, and textures in which there are only displacements alternate. The displacements in the first three gestures alternate between being displaced higher and lower:

5↓ A↑ 3↓
3↑ 0↓
2↑ 5↓ 0B↑

The displacement set emphasizes interval-class five, an "ever extending opposite" of the chromatic pitch-class spaces.

Beaming at the opening of gesture two clearly articulates groups of two. The *forte* dynamic for C#, with continuation of *piano* for F on the fourth eight, breaks apart the association of <15> created in gesture one. <1> is instead associated with the <0B> dyad on the next eight. Likewise, even though <2> on the second eight leads in succession to <3> on the following eight, the beaming makes this association indirect. <01> and <42> are new associations. From this point to the conclusion of the section, the rate at which new incidences are introduced diminishes rapidly. The degree sequence [33332220], begins a trend that evens out figures rather than featuring a disproportionate emphasis.

The cadence of gesture three is derived from the cadence of gesture two, ending at the same pitch-level and reintroducing set-class 6-3, this time partitioned into linear presentations of set-classes 4-11 and 4-7. The degree sequence of this gesture, [33221100], is grouped uniformly into

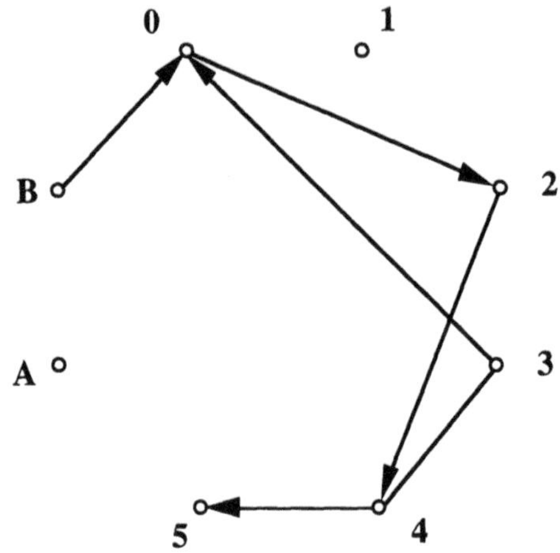

Figure 4-21 Pitch-Class Network: Last 5 Quarters of Gesture Three

pairs of double figures. This pattern associates or relates degrees implicitly by function: <0> and <4> are both degree three, <2> and <3> are degree two, <E> and <5> are degree one, but complement each other FROM and TO. <A> and <1> are degree 0. This degree sequence continues the trend of more equalized figures seen in the previous example. The cycle <02430> is articulated twice in mid-gesture, opposing the placement of cycles in the first gesture at its beginning and ending. The incidence <30> and <45> are new. At the gestural level, the third gesture is no longer expansionary, but largely developmental instead.

The degree sequence for Figure 4-22, [65544332], is still a five-level hierarchy. Compared to the sequence of the first two gestures, [6544 3322], it differs in only three increments.

The fourth gesture is separated from the third by pedal break, 8th rest, the loudest dynamic used to this point, and sudden contrasting rhythms. It immediately expands the highest register by a major third, continuing the registral growth displayed in every gesture. Unlike the first three, gesture four is discontinuous, broken into three fragments.

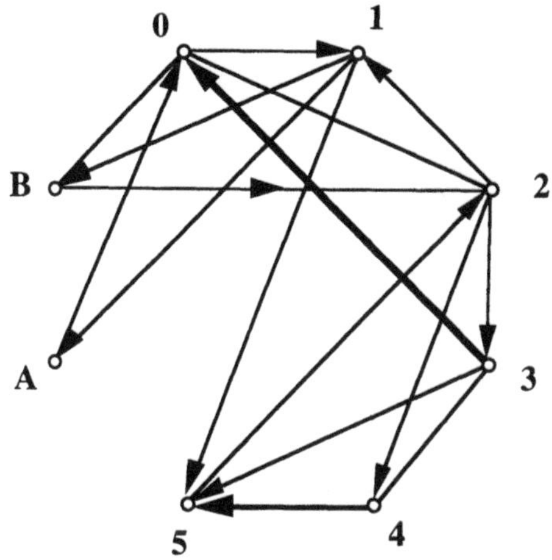

Figure 4-22 Composite Pitch-Class Network: First Three Gestures

The first and third fragments feature large intervals in simultaneous contrary motion. The internal fragment opposes with small intervals in sequential contrary motion grouped into pairs. This internal fragment uses the same pitch-classes from cadential actions in gestures 2 and 3, set-class 6-3 <B02345>. In this context the role of this set-class is reversed, appearing in the middle of a gesture rather than at its cadence. This use of internal reference resembles the cycle from gesture two. The first fragment of gesture four is a broken reference to gesture one, and is the first presentation of all eight pitch-classes since that time. The reappearance of <A> is used to mark this occurrence. The function of <A> throughout is to help mark and articulate statements of the opening octachord.

In Figure 4-23, the incidence <04> is indirect; its association split between gestures in the right and left hands. This same vertical is the first articulated between hands, on the third beat of gesture one. The first vertical in the left-hand of gesture four is the third and strongest association of the <15> dyad, at the same pitch-level as in measure one. The other two associations in gestures one and three are also near their openings. The degree sequence becomes more uniform; [33222222].

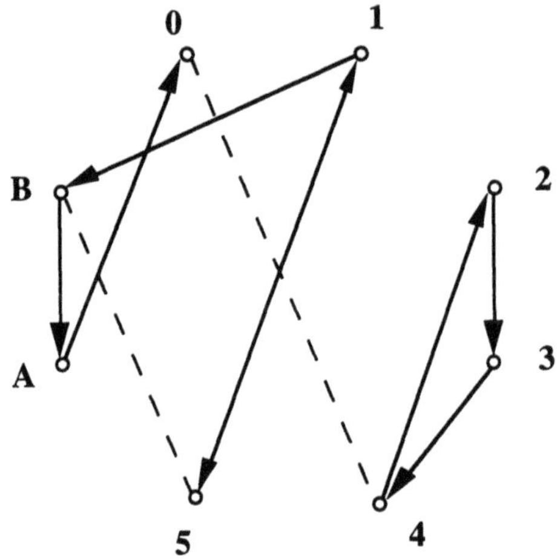

Figure 4-23 Pitch-Class Network: Gesture Four, First Fragment

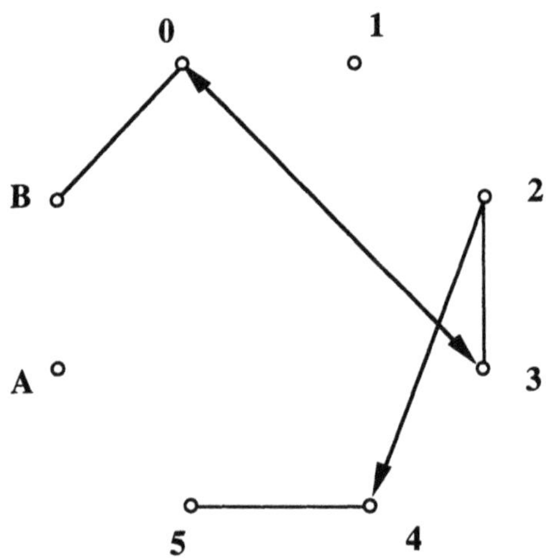

Figure 4-24 Pitch-Class Network: Gesture Four, Second Fragment

The non-Hamiltonian fragment in Figure 4-24 creates a path through associations previously used in cycles at the end of gestures two and three. The left-hand incidence reverses associations made in the right-hand in discontinuous fashion. The degree sequence [22221100] links this fragment to others in its gesture. Comparing it to the cadence of gesture three, one perceives the same 0-incidence, and addition of <03>, <34>, and <54>, while removing the <02>, <34>, and <34> incidence.

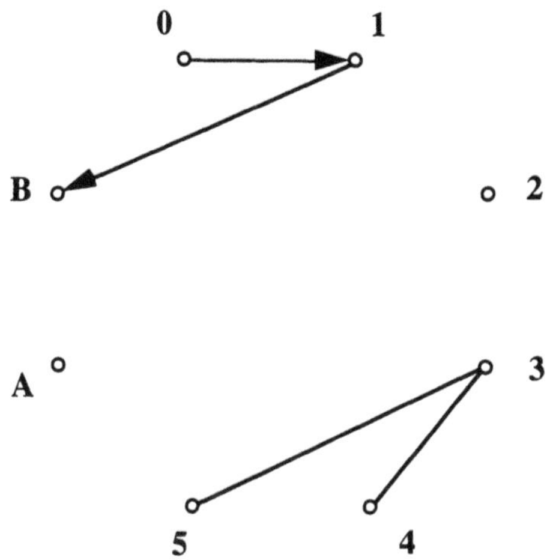

Figure 4-25 Pitch-Class Network: Gesture Four, Third Fragment

Incidence in the third fragment of gesture four is most closely related to the opening of the third gesture. The <1:0B> association made through forte dynamic in gesture three is referenced in the left-hand, gesture four; third fragment. The right-hand of the third fragment is a broken reference to both the dotted association between <4> and the <35> beamed eighths in gesture three, and the <342> cycle at the opening of the fourth gesture. This is a summation of all gestures and like gesture two, it closes with a reference to its opening. This last fragment partitions the vertex set into two parts, each of which has degree [2110]. The incidence pattern of each part is differentiated, in keeping with spirit of "extraordinary, highly layered, highly charged" musical phenomena.

The opening of gesture five is the third complete restatement of the opening octachord at the same pitch-level, reordered to introduce several new incidence patterns.

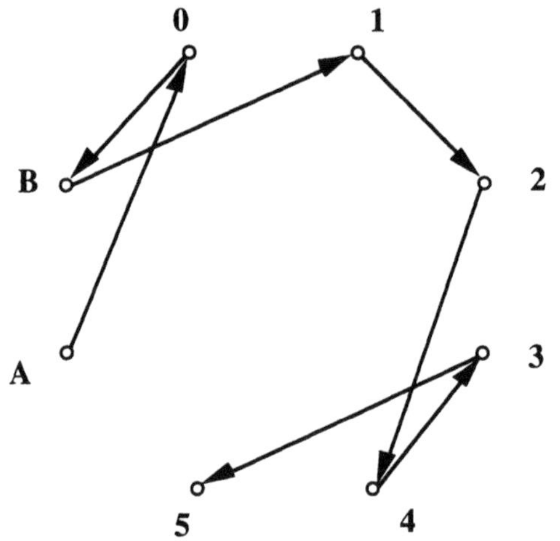

Figure 4-26 Pitch-Class Network: Gesture Five, First 8 Eighth Notes

This Hamiltonian path cadences on the same pitch as the second and third gestures. Pitch-class <A>, present again in a complete statement of the opening octachord, is used for the first time as an opening pitch. <A0B> is the only incidence kept from the opening gesture, all other connections differ. <B1> and <12> are connected for the first time in the piece, and <1> is presented in a previously unused register to help create a new sense of linearity in this gesture.

Comparing the opening of gesture five to the first fragment of gesture four; where the complete octachord is also sounded, an association complex is formed that reveals the extent to which differentiation of articulated patterns is used to propel development of musical thoughts. The association <A0> is the only one shared between them. <B1>, <24>, and <43> appear in both but reversed in directional association. Gesture five contains three associations not in gesture four first fragment; <0B>, <12> and <35>. The patterning of reversals and differences alternates after the first association, until the <43> and <35> association, at which time a reversal occurs; leaving two reversals together concluding with a difference in association.

<A0> Contained in both gestures four-A and gesture five
<0B> Not contained in gesture four-A
<B1> Reversed between gestures four-A and five
<12> Not contained in gesture four-A
<24> Reversed between gestures four-A and five
<43> Reversed between gestures four-A and five
<35> Not contained in gesture four-A

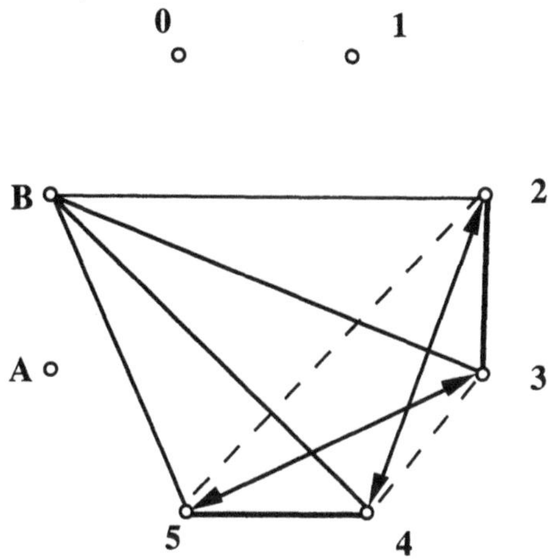

Figure 4-27 Pitch-Class Network: Cadence of Gesture Five

The cadence of gesture five uses a voice exchange (notated in bold lines) between dyads <24> and <35>, (notated as verticals using double arrows) which are also associated by beaming at the opening of gesture three. The association between <B4> is fresh, and <B5> is an indirect association at one other point; Figure 4-23.

Gesture six extends the upper and lower pitch boundaries again, its first three pitch-classes sounding in new octaves. It reintroduces the eight primary pitch-classes; the last time before the four unused pitch-classes that complete the aggregate are introduced. Several new incidences are saved for use at this time, including <54>, <40>, <0A>, <13>, and <2B>.

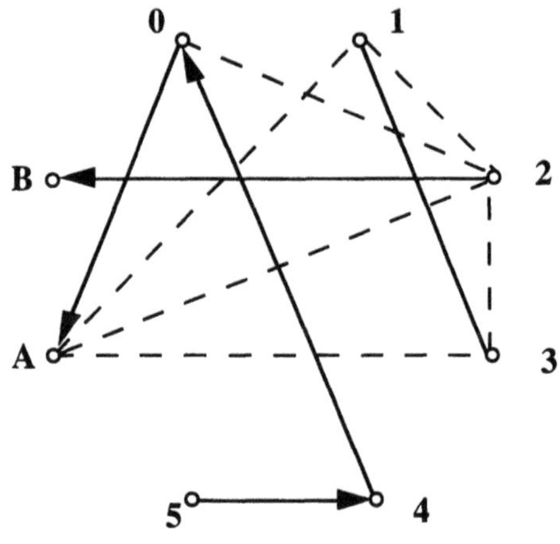

Figure 4-28 Pitch-Class Network: First 6 Eighths of Gesture Six

Three of the four missing pitch-classes are introduced in gesture six, as well as a collection of new incidence patterns.

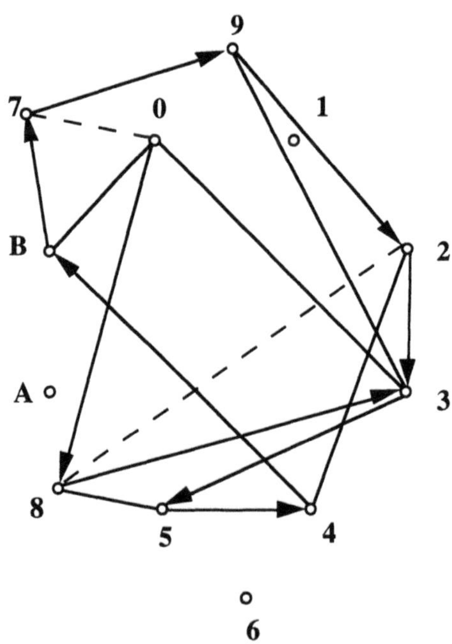

Figure 4-29 Pitch-Class Network: Cadence of Gesture Six

The cadence of gesture six is a variation and extension of the cadence of gesture three, which also leaves <A> and <1> 0-incident. The first two new pitch-classes introduced in this passage, <7> and <9>, are sounded against the <0B> dyad heard prominently in gesture three. The cycle <24B0> , set-class 4-11, is derived from the <B024> cycle from gesture three, making the association between the cadence of gesture three and gesture seven stronger. Pitch-class <8> is more strongly integrated into the texture than are any of the other complementary pitch-classes, immediately establishing a degree of four. Again, new associations are saved to make this gesture fresh, these including <4B>, <39>, and <92>.

The degree sequence can now include all twelve positions in the aggregate, [544444333000]. This new level-4 sequence continues the overall incidence strategy, with degree sequences become more uniform, and emphasis shifting toward pitch-classes other than those highlighted at the opening. Pitch-class <3>, not <2>, is the most connected degree in this passage.

Figure 4-30 Gesture Seven

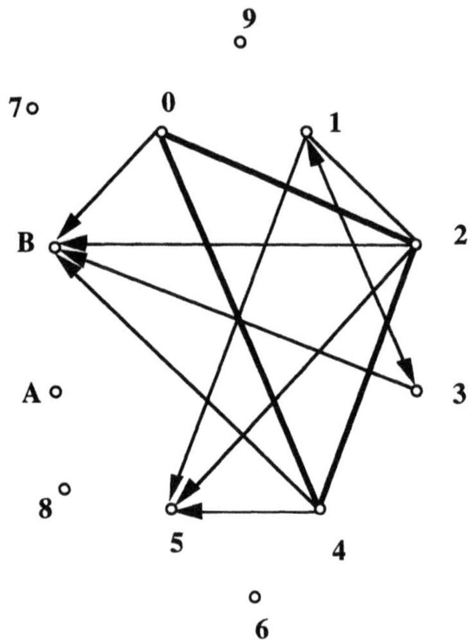

Figure 4-31 Pitch-Class Network: Gesture Seven, First Five Triplet Quarters

The opening of gesture seven contains two events related by literal pitch-space inversion around D above middle C. The degree sequence reinforces this symmetry [66664440000], the central collection <024> receiving greater emphasis. This manifestation of symmetry around D is a surface articulation that resonates with the atomic prevalence of <2> in the opening gestures. The same pitch is used for the sudden cadence of gesture four fragment one.

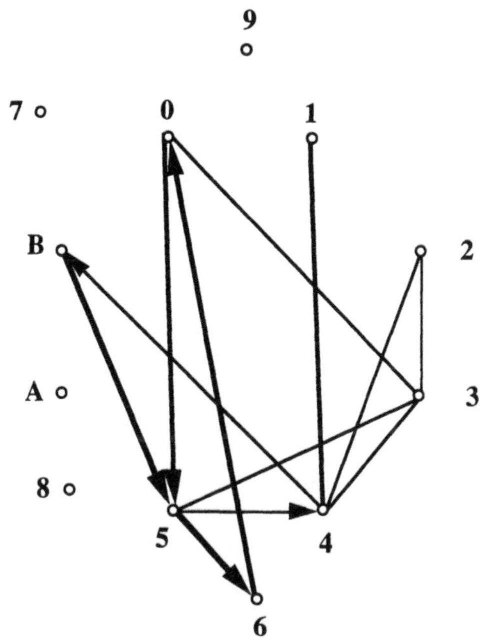

Figure 4-32 Pitch-Class Network: Cadence of Gesture Seven

Both hands in the cadence of gesture seven contain pitch-class set 6-3. The right-hand, <B02345>, is the same pitch material as from the cadence of gesture two, the cadence of gesture three, and the internal fragment of gesture four. The transformation of set-class 6-3 in the left-hand, <013456>, is the retrograde of the right-hand transposed one semitone higher. The new association <B5> is framed in the center of a four-element pitch-class retrograde on either side of it: <0324:B5:4230>. The rare use of a standard transformation, is also used in the left-hand <1435:60:5341> to introduce <6>, completing the aggregate.

The degree sequence [554322210000] is six-level, and like gesture six emphasizes <5> and <3>. Several new associations are saved for this last gesture within this larger section, <B5>, <05>, <14>, and <41> are new. As in its presentation in gesture five, <1> is presented in a new octave, saving a register for closure and freshness.

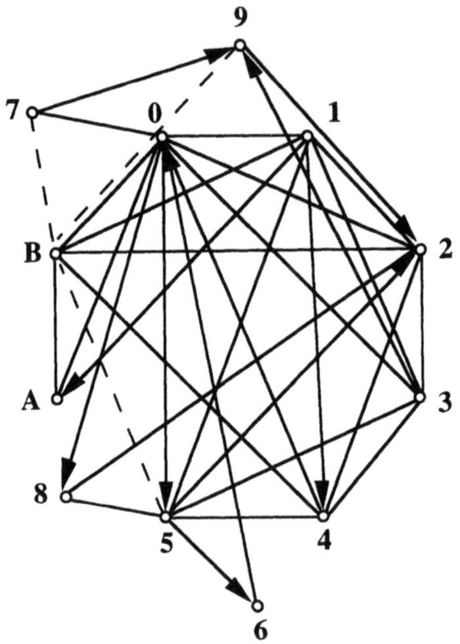

Figure 4-33 Composite Pitch-Class Network of Entire Section

In Figure 4-33, the composite degree sequence [B88776643332], is a 7-level cyclic hierarchy. Complementary pitch-classes are sounded without being particularly assimilated into the overall complex. This marks them as foreign sounding events and emphasizes their presence as an element of aggregate completion. At the conclusion of *Form IV*'s first complete section, all fourteen possible interval-class one, and all twelve possible interval-class two associations are used. Patterns of association involving pitch-classes from within the opening octachord that remain unused include three interval-class three associations: <A1>, <14>, and <25>. Five of eight possible interval-class fours are not used: <04:2A>, <3B>, <B3>, and <51>. Five of eight possible interval-class fives are not used: <3A>, <A3>, <50>, <5A>, and <A5>. Two of four possible interval-class six associations are not used <4A> and <A4>. Pitch-class <A> functions like a mirror of the complementary tones within the opening octachord. Its presence marks all statements where the entire octachord is presented, yet its degree and number of occurrences are similar only to members of the complementary tetrachord. Pitch-classes <1>, <4>, and <A> are graphically distinct in that they connect only inside the opening octachord. Pitch-class <1> is prominent in both the opening and closing gestures of the section, and is realized as the lowest pitch. It has half the number of occurrences of most other pitch-classes in the octachord, excluding <A>. Pitch-class <4> opens the second and third gestures, figures prominently in gestures where the opening octachord is not complete. The only pitch-class to which <4> is not adjacent in the opening octachord is <1>.

It is hoped that structural potentials inherent in awareness of graphic models of abstract musical transformations will attract the curiosity of a wide range of composers and theorists, and be applied to an equally wide range of styles. It is hoped that these models are further developed and explored as symbols and as signifiers. Adaptable where relations between pitches are important; they offer the potential for creation of compelling musical statements and developmental relations. It is also my hope that these transformations resonate with new aesthetics: to suggest to, and interact with, the compositional imagination.

Exercises

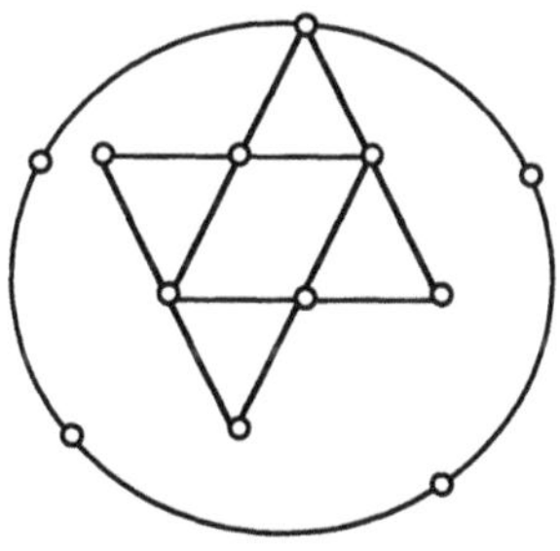

Figure 5-1

1. What is the correct degree sequence of Figure 5-1?

2. Assign pitch-classes to Figure 5-1 in three ways that bring out properties as distinctly different from one another as you are able to make them. Analyze each network.

3. Compose a short realization of Figure 5-1 that makes audible several compelling aspects of it.

4. What is the complementary cycle-class in Figure 5-1?

5. Create a graph with the same degree sequence as Figure 5-1 that is nonisomorphic to it.

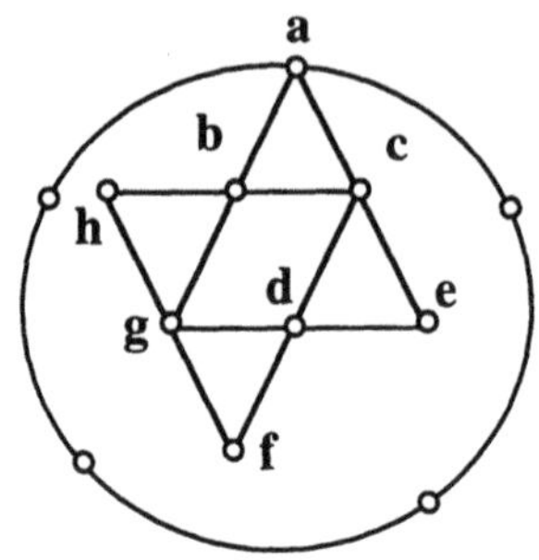

Figure 5-2

6. There are four intersecting trichordal cycle-classes inside Figure 5-2: [abc], [cde], [gdf], [ghb]. All four can be realized with the same set-class using which of the twelve trichordal set-classes? Which set-classes are impossible to realize in this way?

7. List all configurations in Figure 5-2 where all four trichordal set-classes be realized with different set-classes for each trichordal cycle-class?

8. Draw a Hamiltonian and a non-Hamiltonian graph on eight vertices. Assign pitch-classes to both graphs in two ways that bring out properties as distinctly different from one another as you are able to make them. Analyze each.

9. Choose two complementary hexachordal graphs from Appendix I and assign pitch-classes to them in a way that brings out a property of interest. Analyze this property. Compose a realization of the network for small chamber ensemble.

10. Create as many nonisomorphic graphs as is possible from the level-five degree sequence [544333222211].

11. Assign pitch-classes to Figure 3-24 in a way that brings out both pentachordal and hexachordal properties of interest. Describe this property. Compose a short realization of the network for piano.

12. Assign pitch-classes to Figure 3-9 in a way that brings out hexachordal properties of interest. Describe this property. Compose a short orchestral realization of the network.

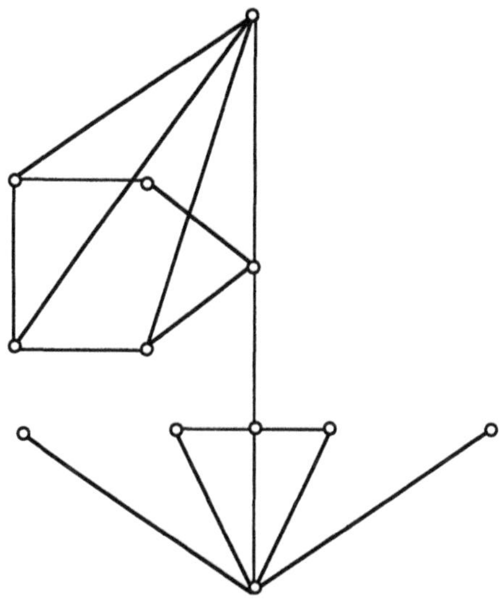

Figure 5-3

13. What is the correct degree sequence of Figure 5-3, and what type of hierarchy would this graph be classified as?

14. Write-out all non-reflexive cycle-classes in Figure 5-3?

15. Assign pitch-classes to Figure 5-3 in three ways that bring out properties as distinctly different from one another as you are able to make them. Analyze each network.

16. Compose a short realization of Figure 5-3 that makes audible several compelling aspects of it.

Appendix I

COMPLEMENTARY HEXACHORDAL GRAPHS

E:15 L:1 [555555]

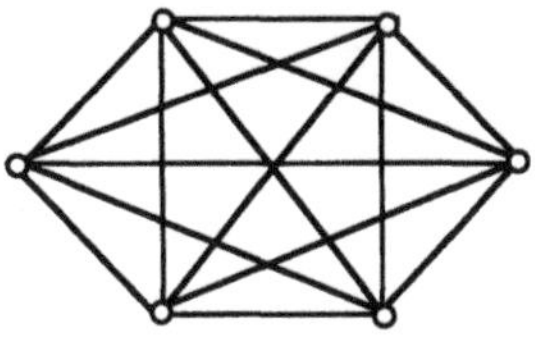

E:0 L:1 [000000]

E:14 L:2 [555544]

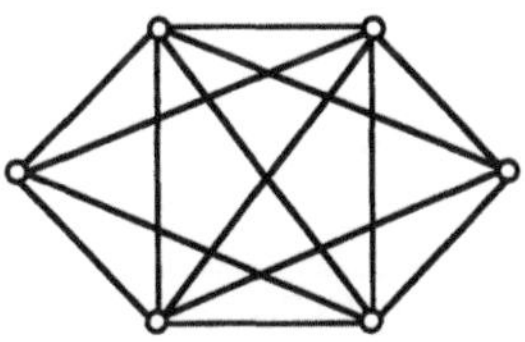

E:1 L:2 [110000]

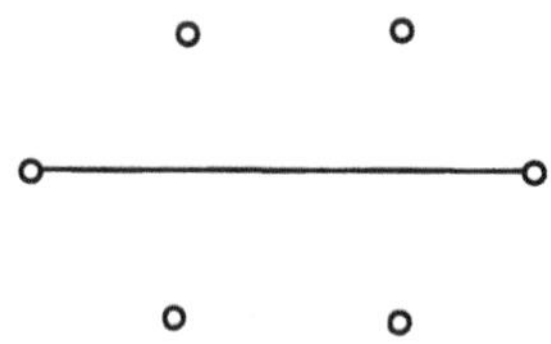

E:13 L:3 [555443]

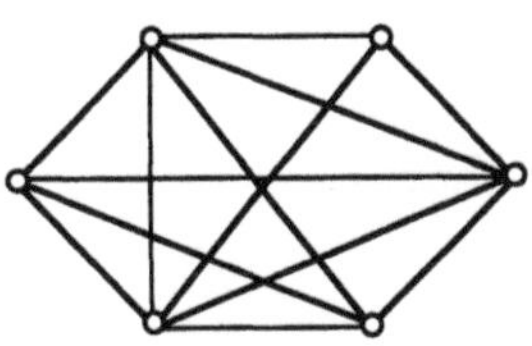

E:2 L:3 [211000]

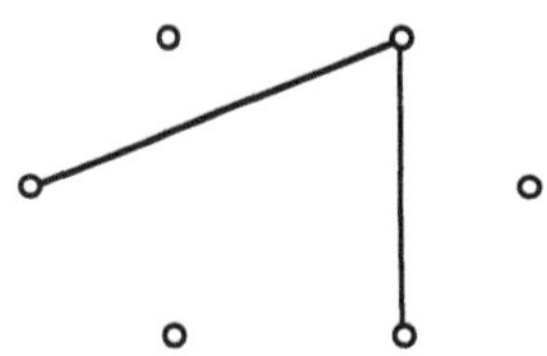

E:13 L:2 [554444]

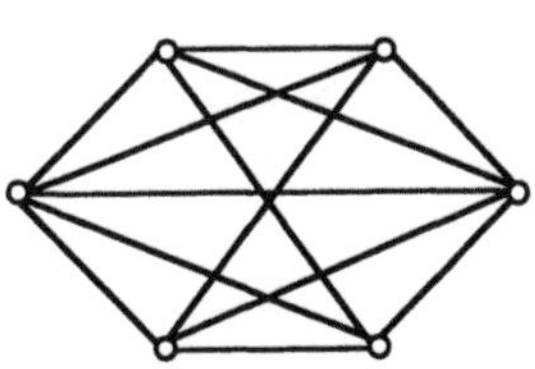

E:2 L:2 [111100]

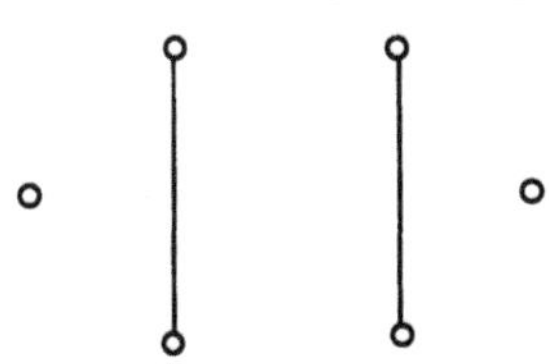

E:12 L:2 [555333]

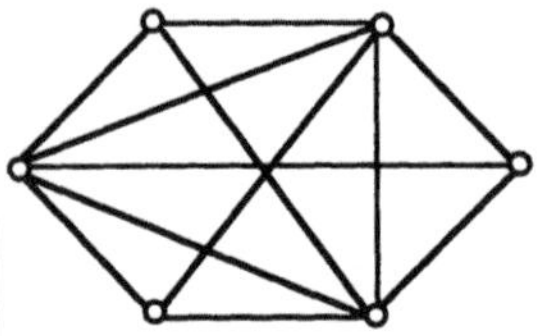

E:3 L:2 [222000]

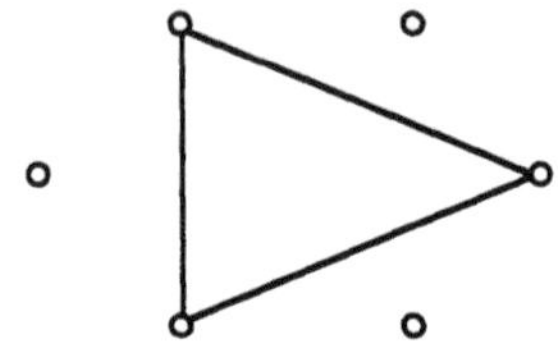

E:12 L:3 [554442]

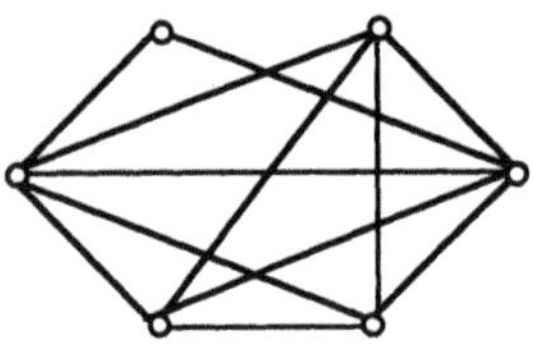

E:3 L:3 [311100]

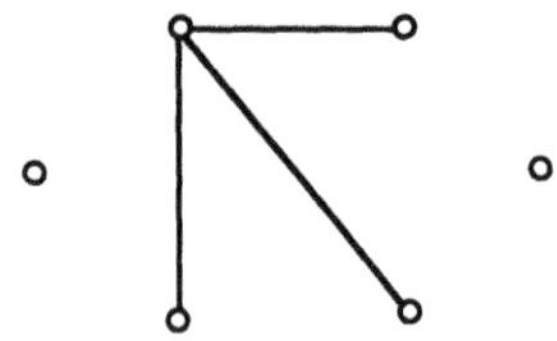

E:12 L:3 [554433]

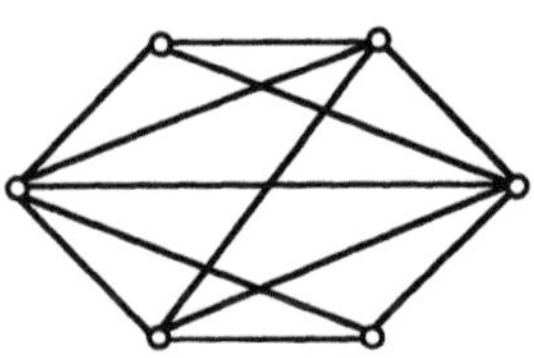

E:3 L:3 [221100]

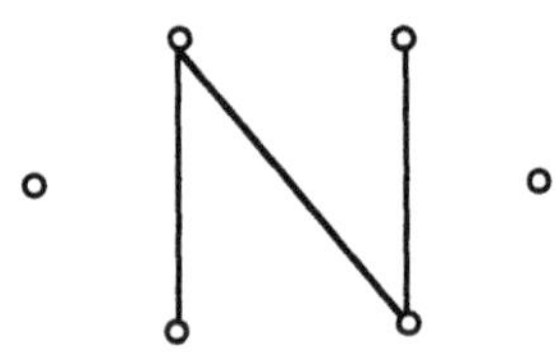

E:12 L:3 [544443]

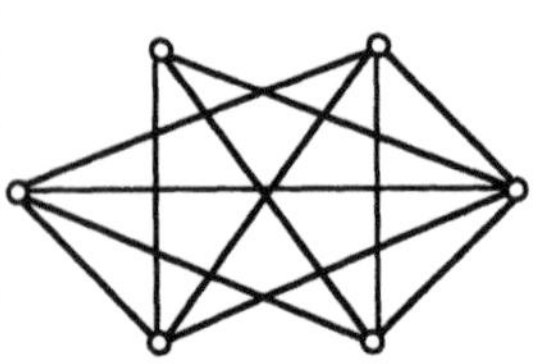

E:3 L:3 [211110]

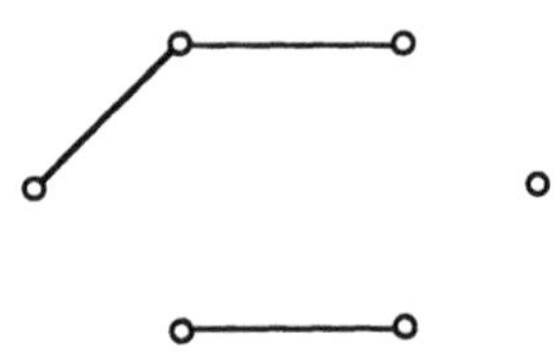

E:12 L:1 [444444]

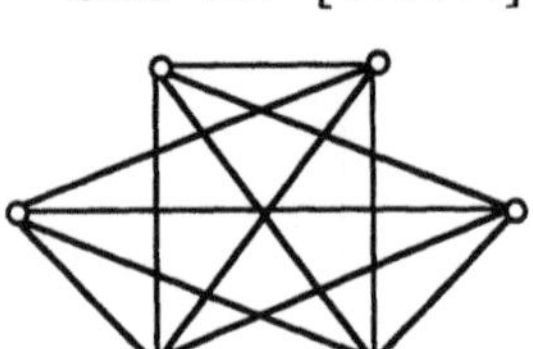

E:3 L:1 [111111]

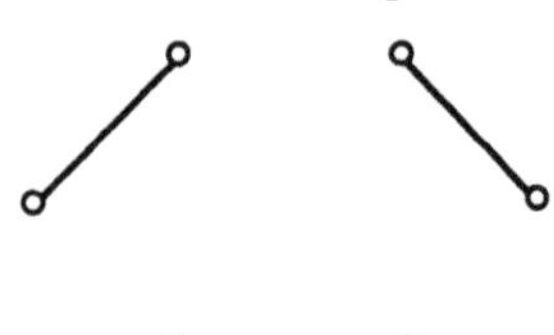

E:11 L:4 [554332]

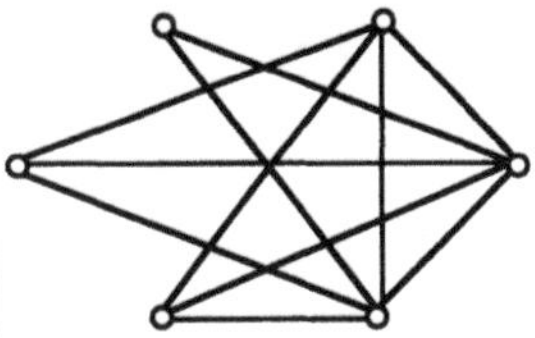

E:4 L:4 [322100]

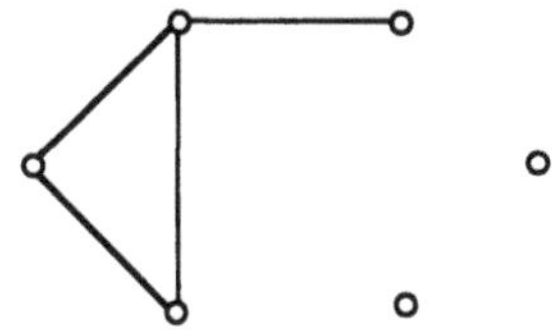

E:11 L:2 [553333]

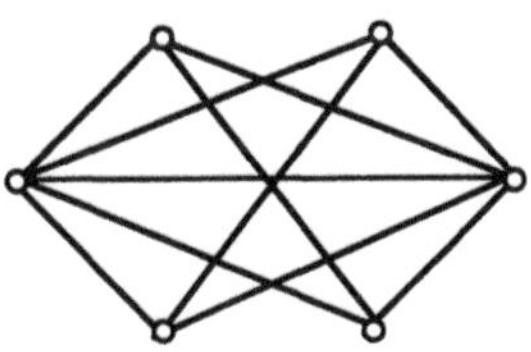

E:4 L:2 [222200]

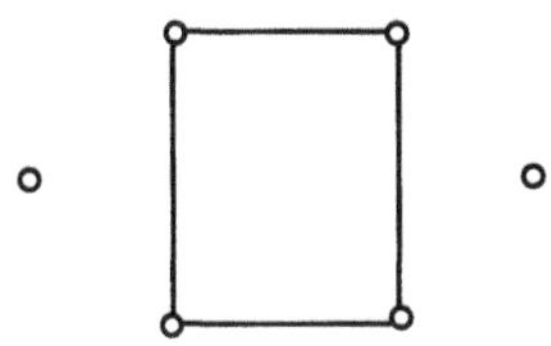

E:11 L:3 [544441]

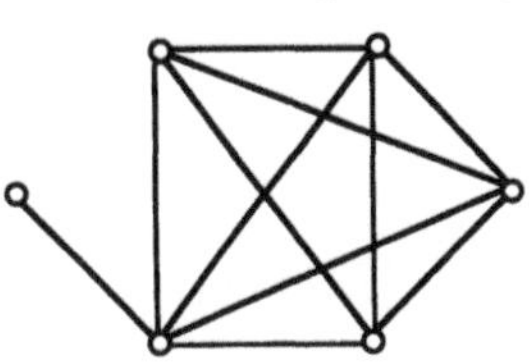

E:4 L:3 [411110]

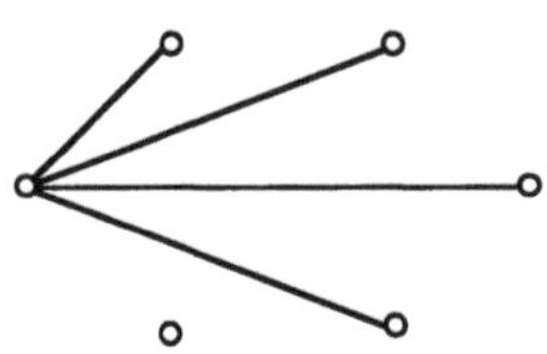

E:11 L:4 [544432]

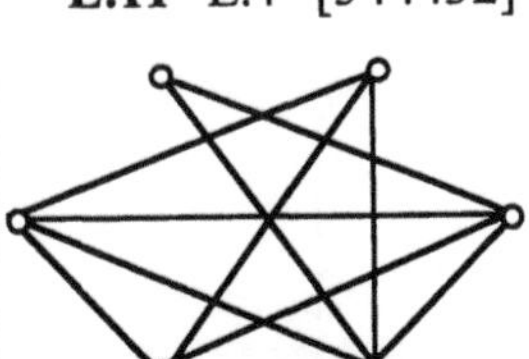

E:4 L4 [321110]

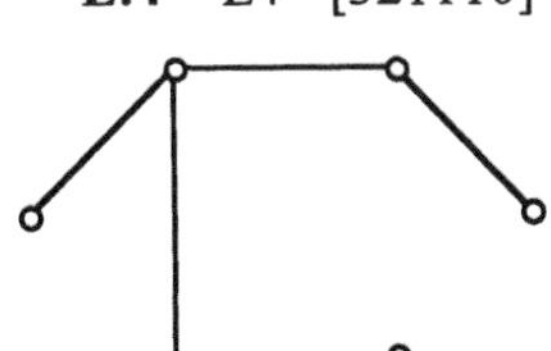

E:11 L:3 [544333]

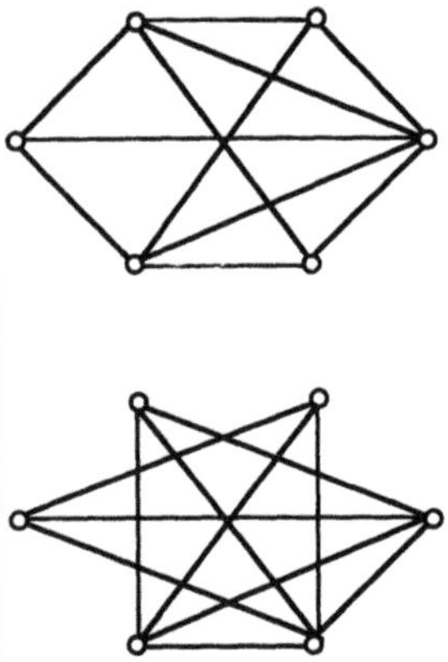

E:4 L:3 [222110]

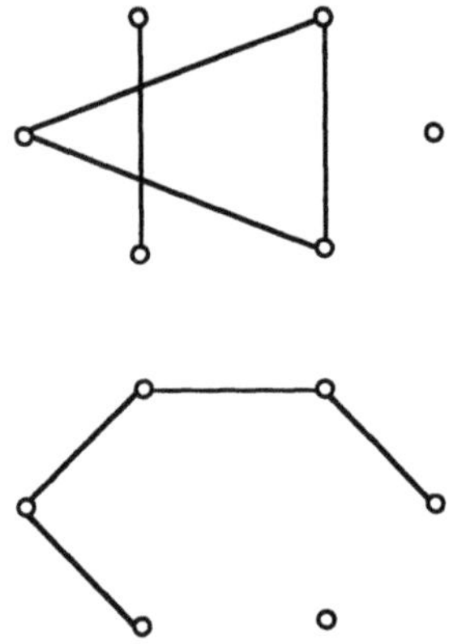

E:11 L:2 [444442]

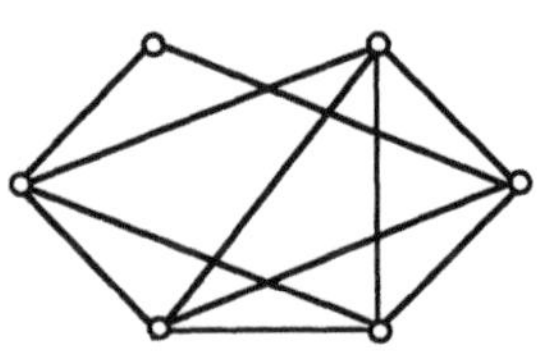

E:4 L:2 [311111]

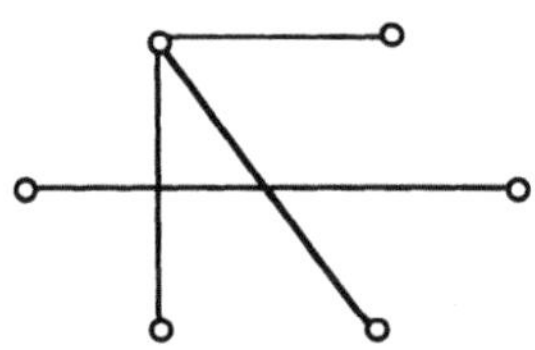

E:11 L:2 [444433]

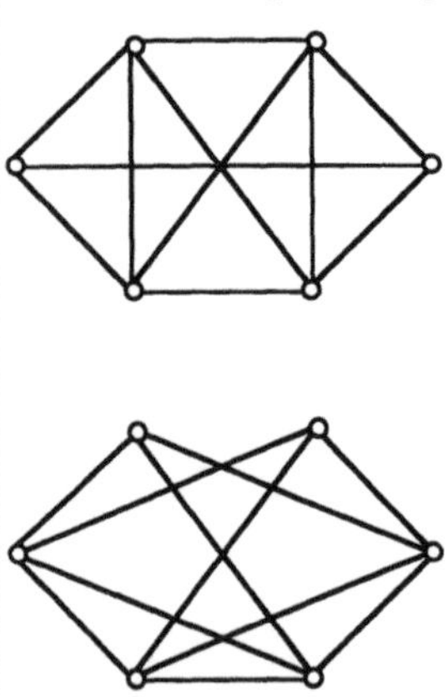

E:4 L:2 [221111]

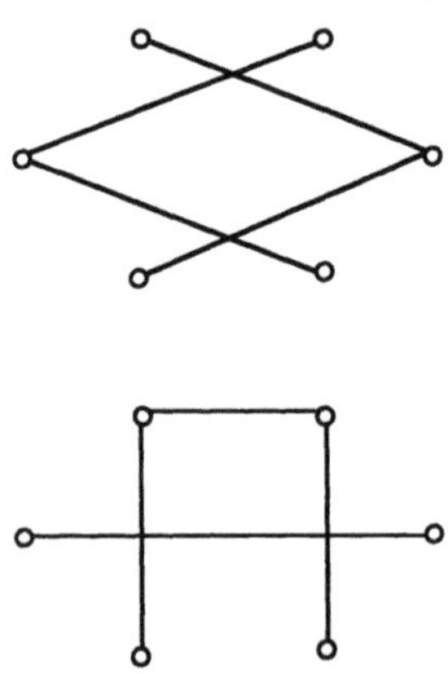

E:10 L:3 [553322]

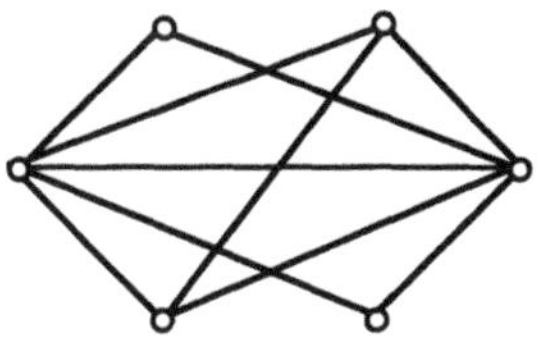

E:5 L:3 [332200]

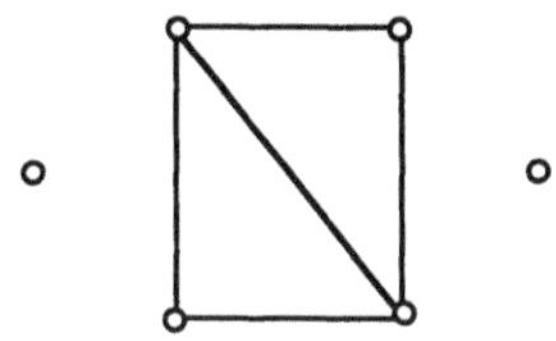

E:10 L:4 [544331]

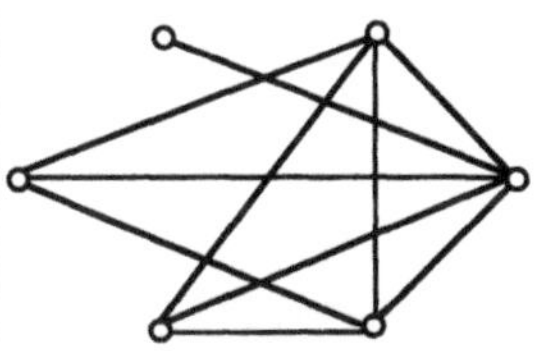

E:5 L:4 [422110]

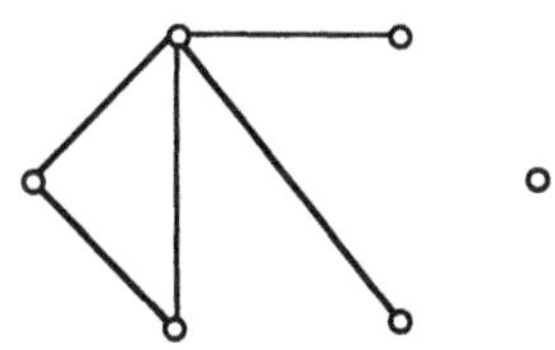

E:10 L:4 [544322]

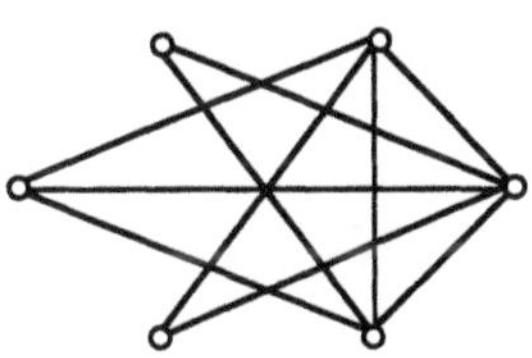

E:5 L:4 [332110]

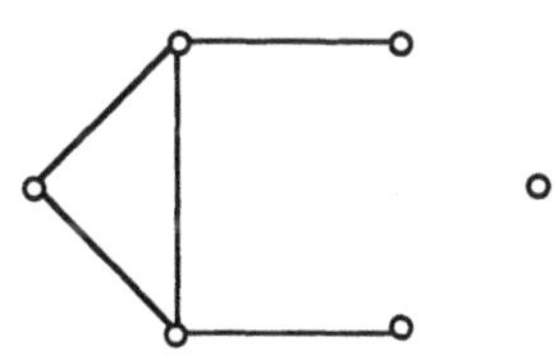

E:10 L:4 [543332]

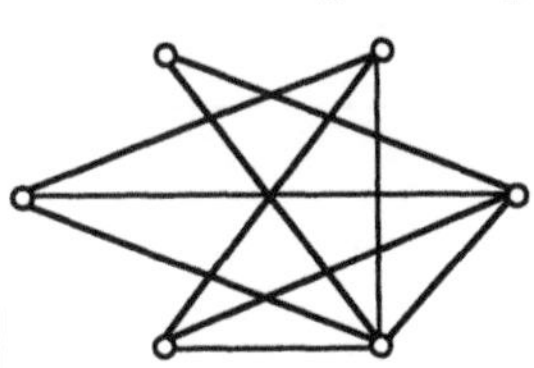

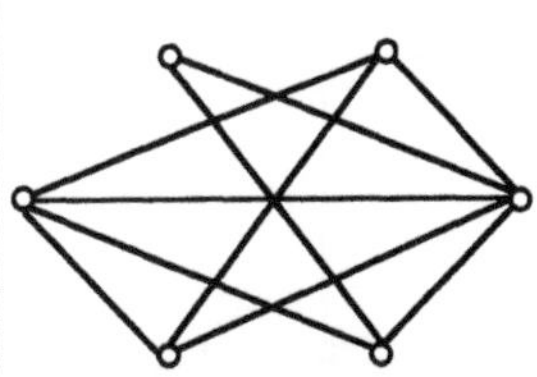

E:5 L:4 [322210]

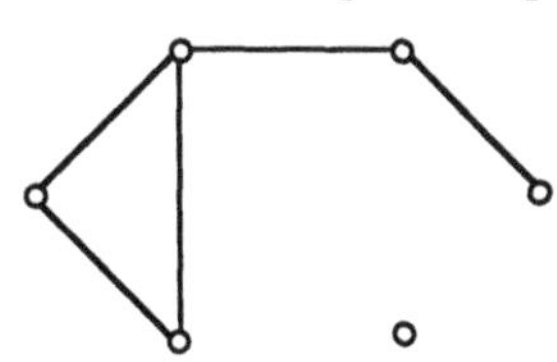

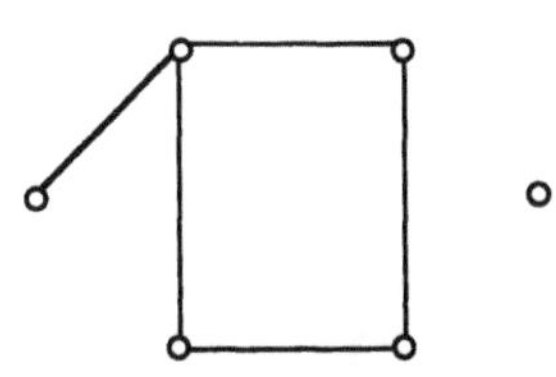

E:10 L:2 [533333]

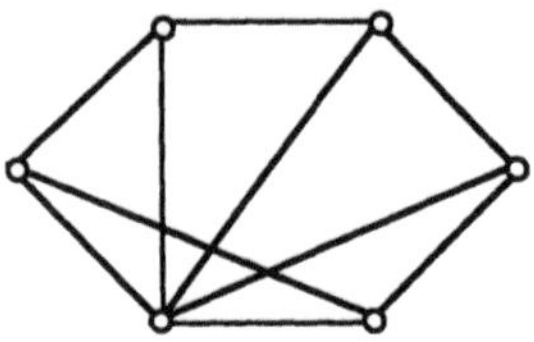

E:5 L:2 [222220]

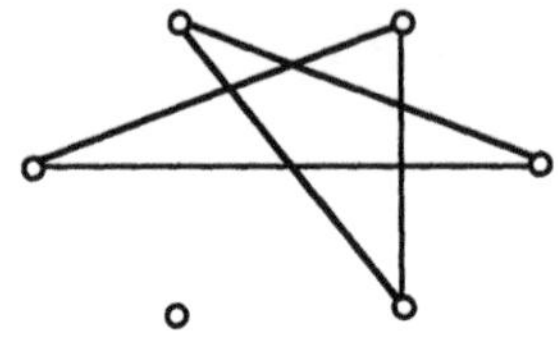

E:10 L:2 [444440]

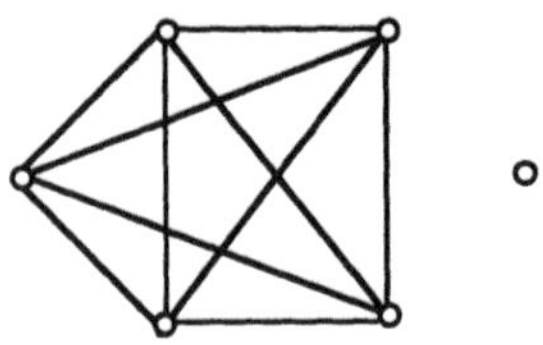

E:5 L:2 [511111]

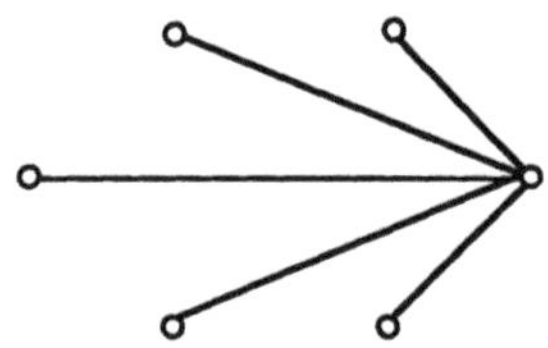

E:10 L:3 [444431]

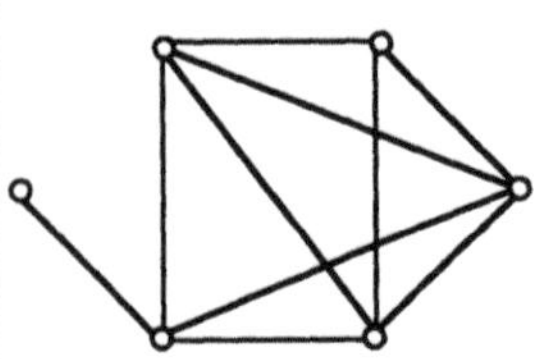

E:5 L:3 [421111]

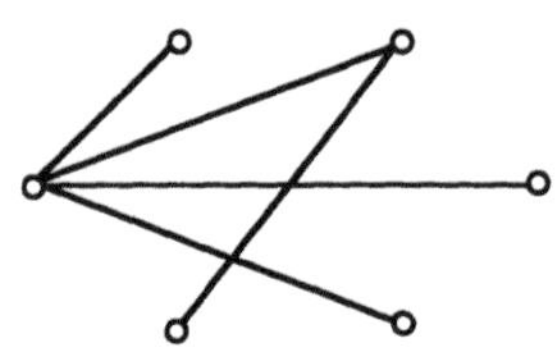

E:10 L:2 [444422]

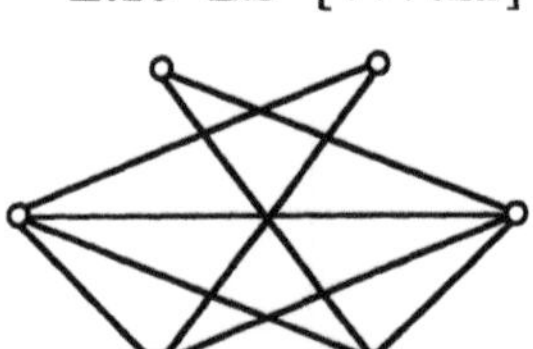

E:5 L:2 [331111]

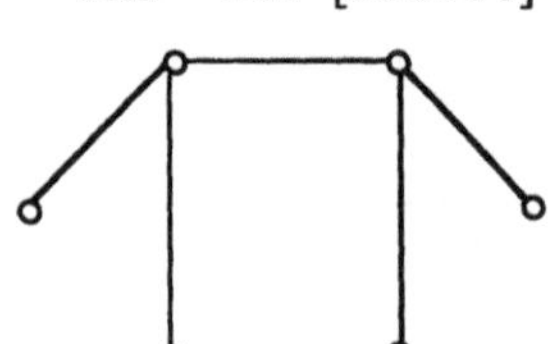

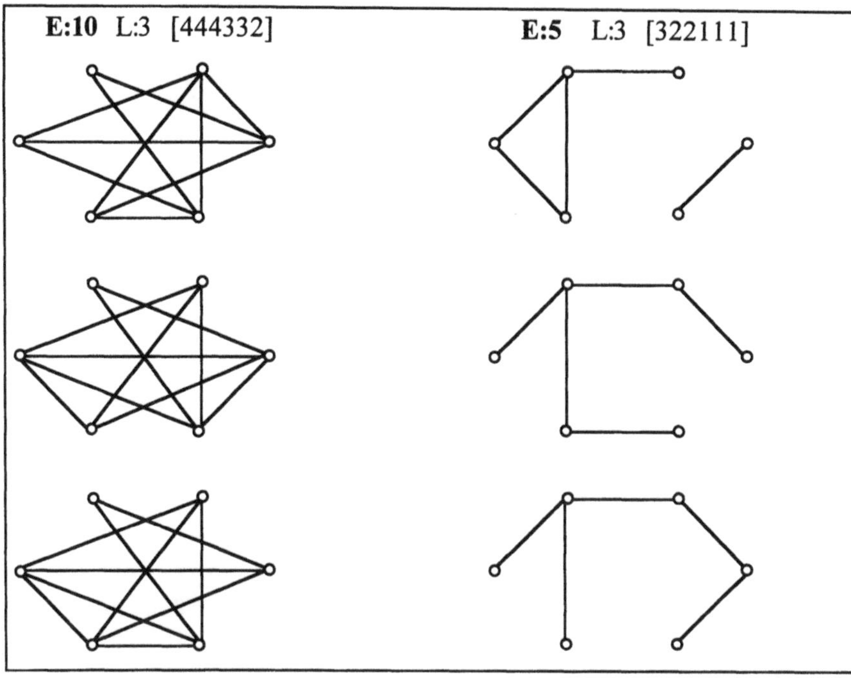
E:10 L:3 [444332]
E:5 L:3 [322111]

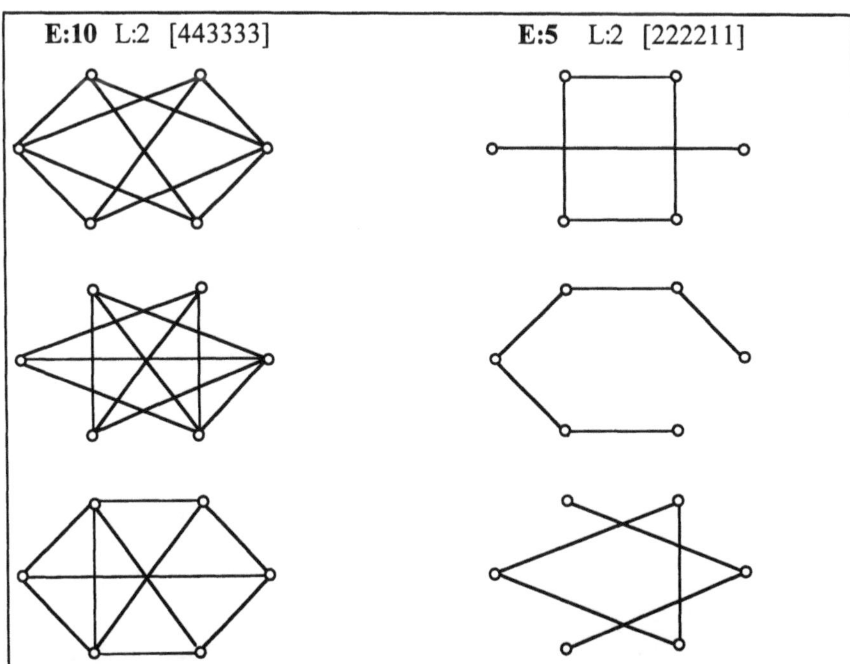
E:10 L:2 [443333]
E:5 L:2 [222211]

E:9 L:2 [552222]

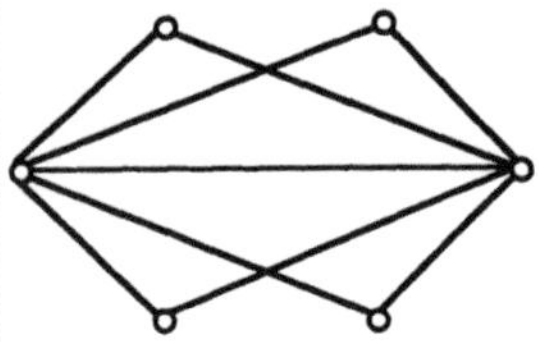

E:6 L:2 [333300]

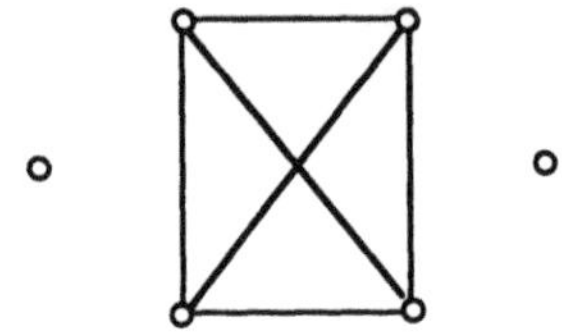

E:9 L:5 [543321]

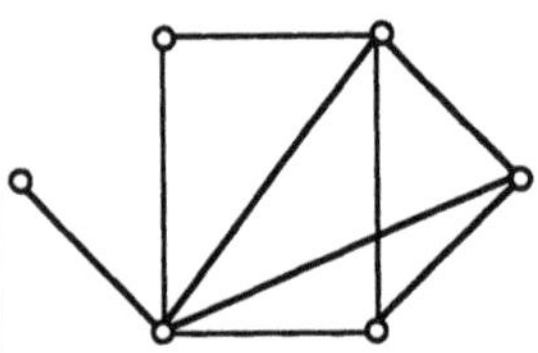

E:6 L:5 [432210]

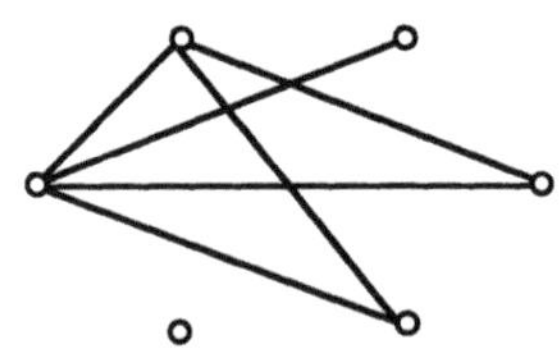

E:9 L:4 [543222]

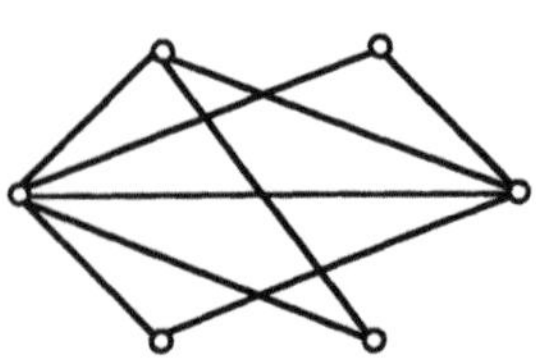

E:6 L:4 [333210]

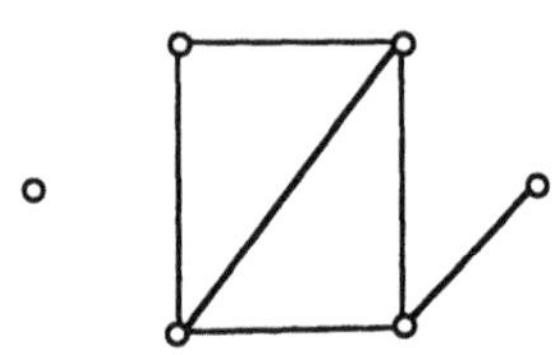

E:9 L:3 [533331]

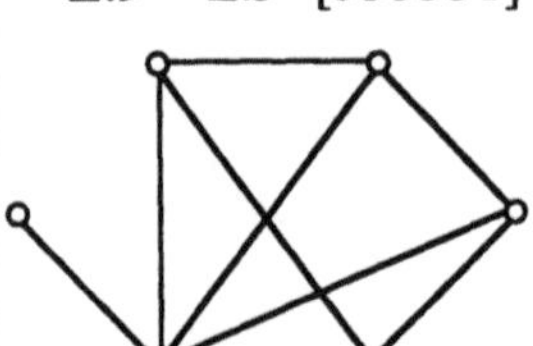

E:6 L:3 [422220]

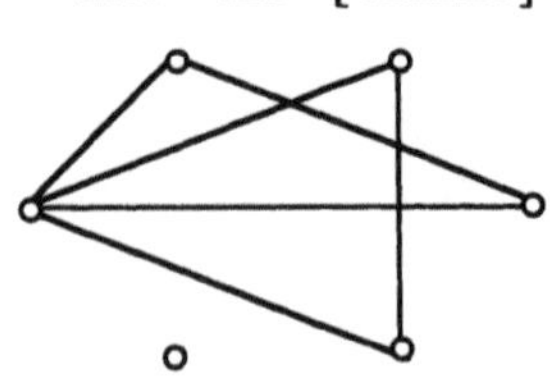

E:9 L:3 [533322]

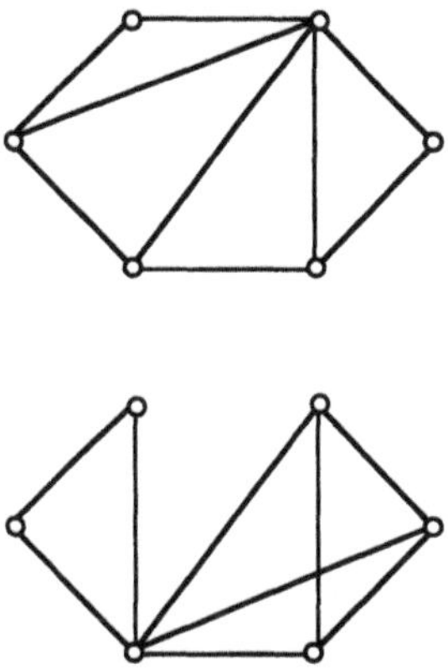

E:6 L:3 [332220]

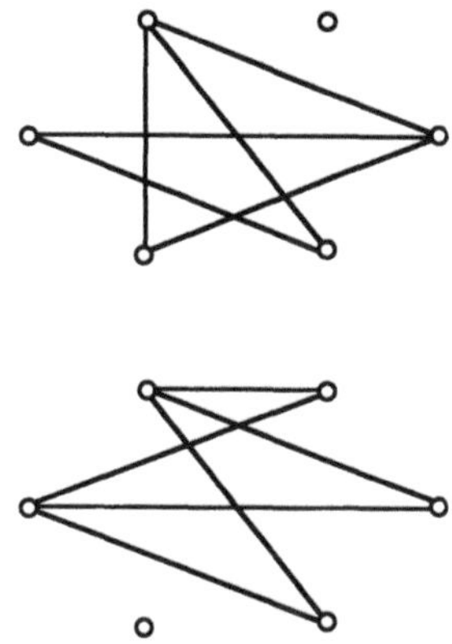

E:9 L:3 [444330]

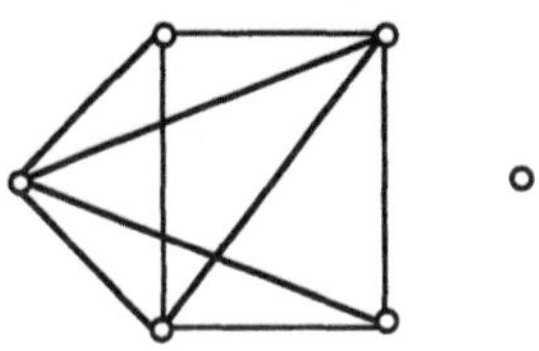

E:6 L:3 [522111]

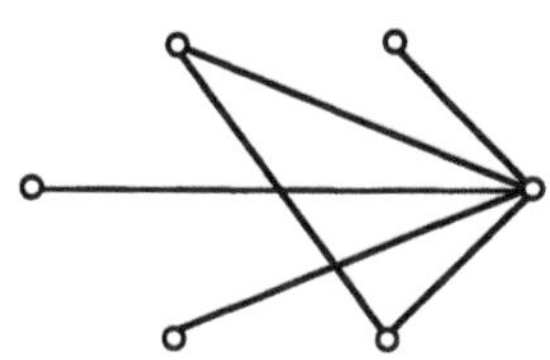

E:9 L:4 [444321]

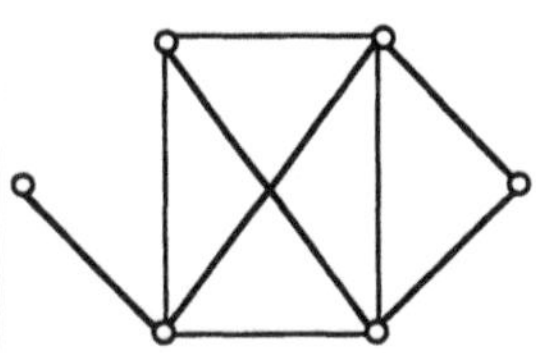

E:6 L:4 [432111]

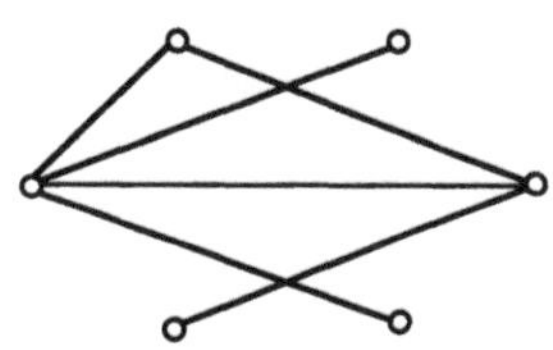

E:9 L:2 [444222]

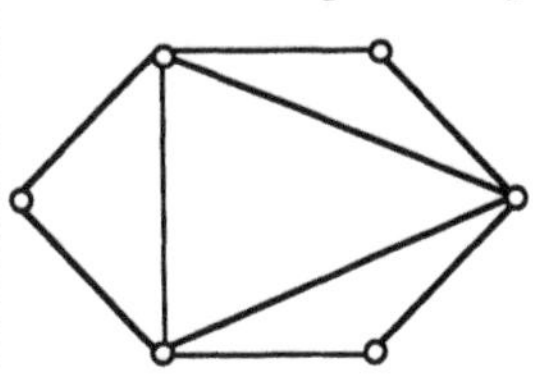

E:6 L:2 [333111]

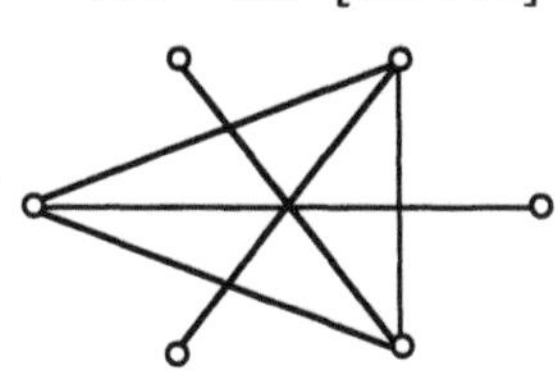

E:9 L:3 [443322]

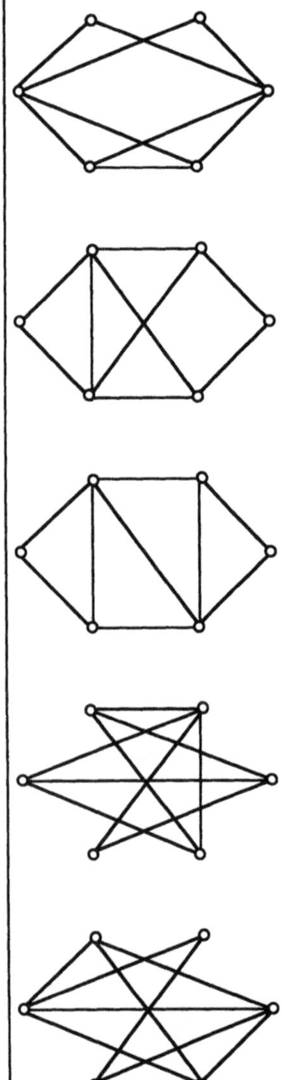

E:6 L:3 [332211]

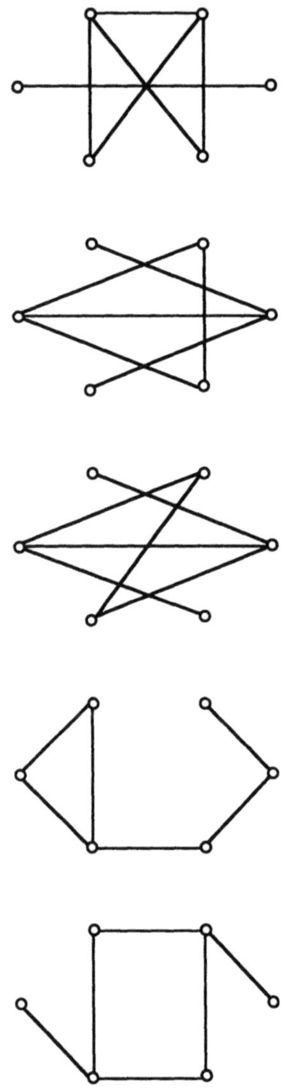

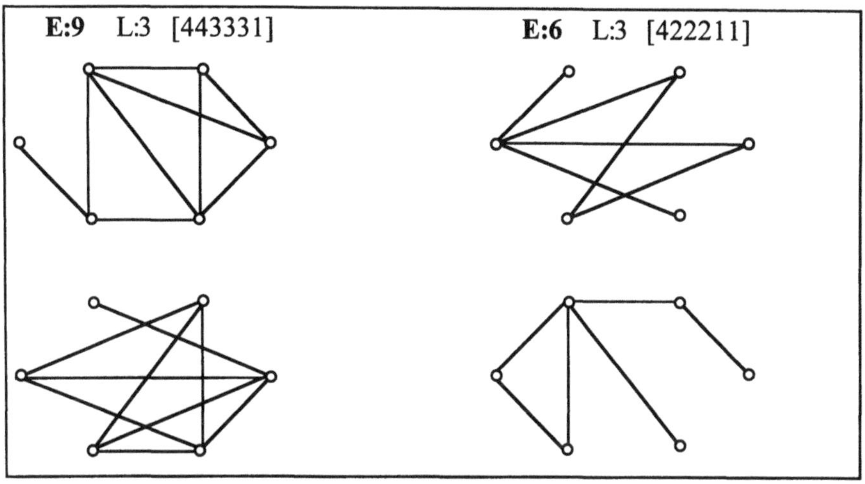
E:9 L:3 [443331]
E:6 L:3 [422211]

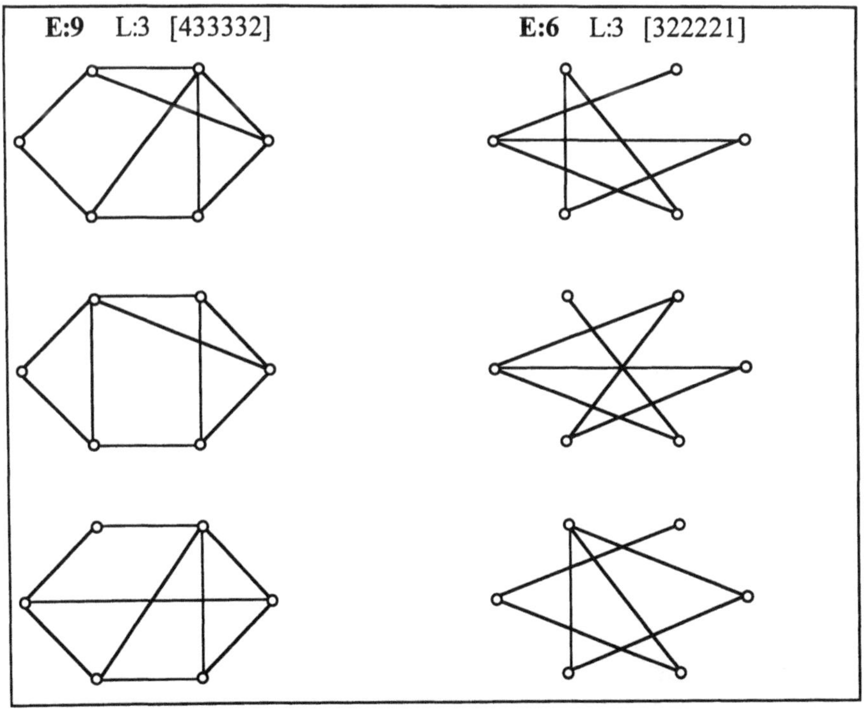
E:9 L:3 [433332]
E:6 L:3 [322221]

E:9 L:1 [333333]

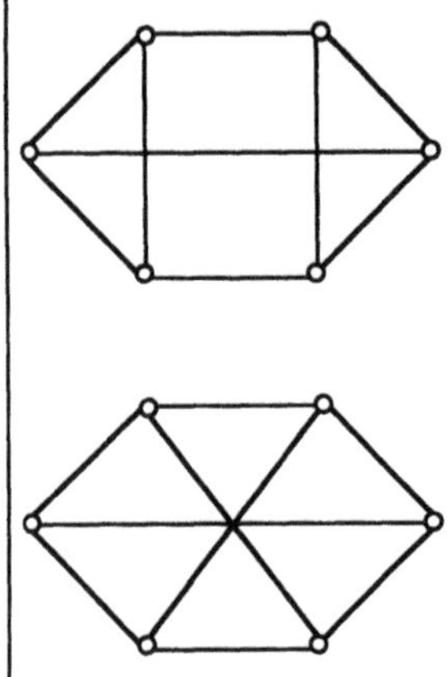

E:6 L:1 [222222]

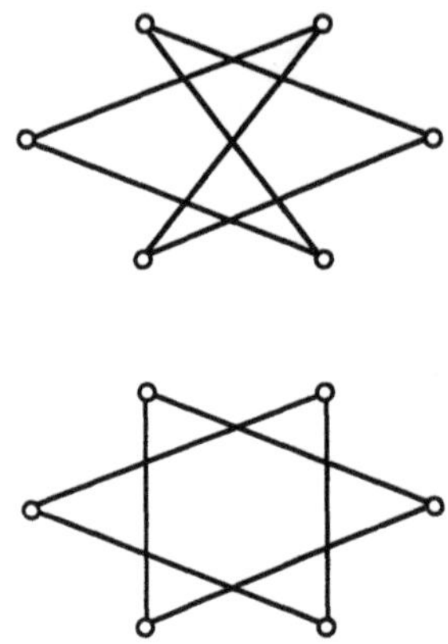

E:8 L:4 [542221]

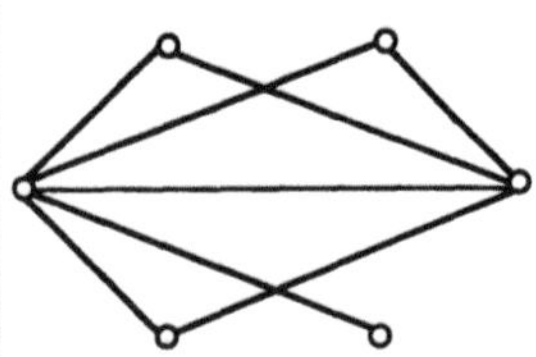

E:7 L:4 [433310]

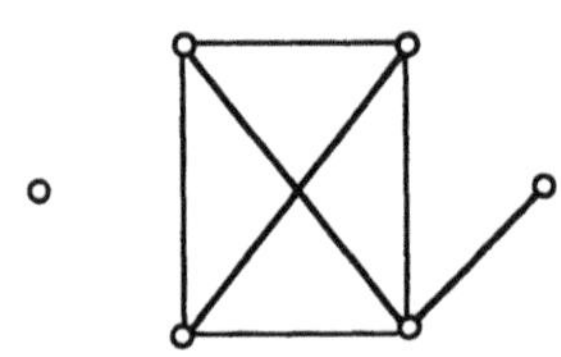

E:8 L:3 [533311]

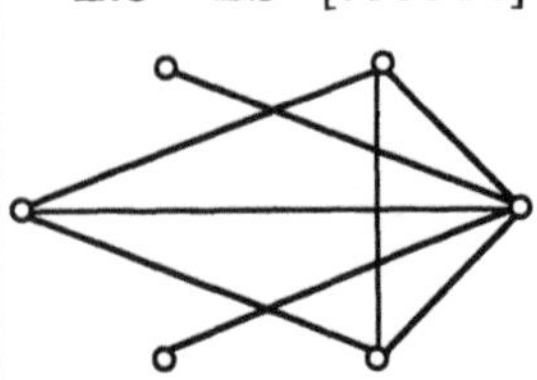

E:7 L:3 [442220]

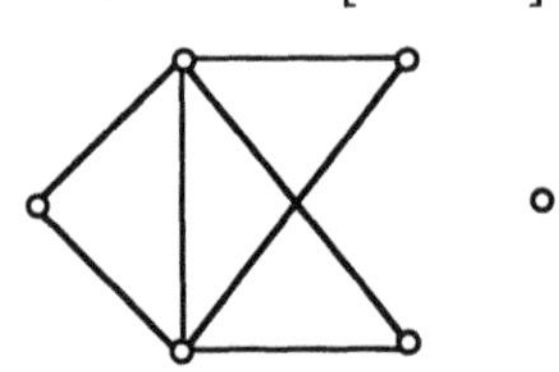

E:8 L:4 [533221]

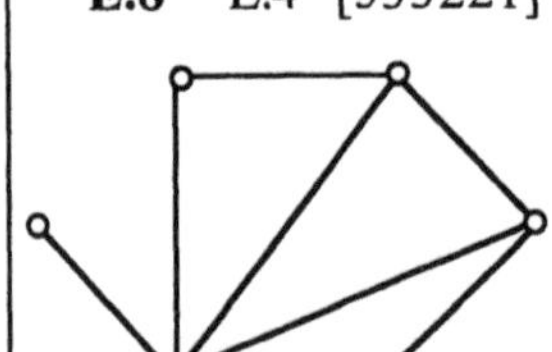

E:7 L:4 [433220]

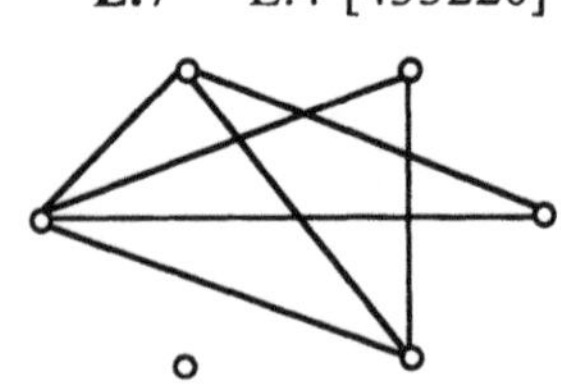

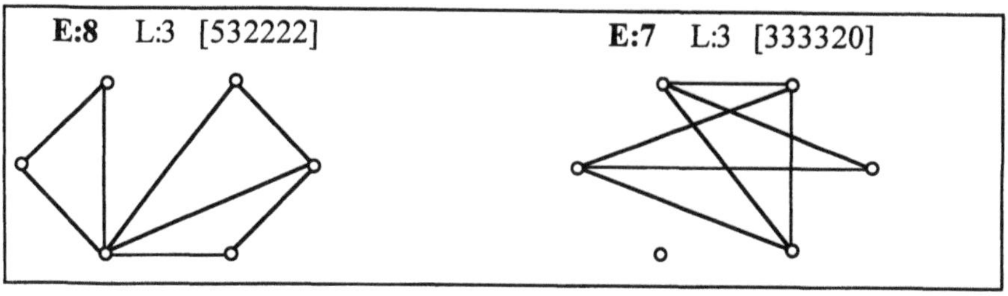
E:8 L:3 [532222]
E:7 L:3 [333320]

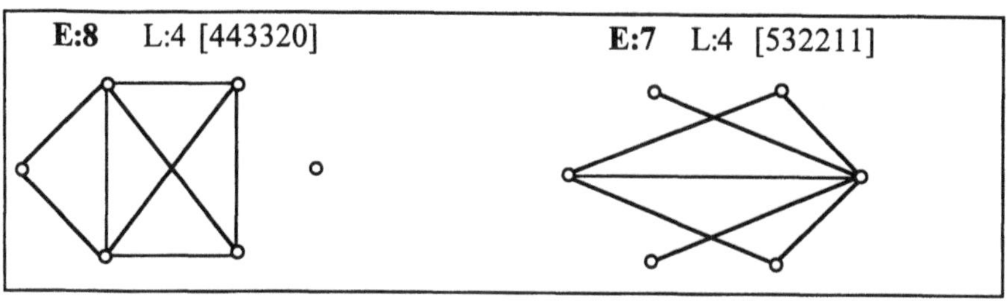
E:8 L:4 [443320]
E:7 L:4 [532211]

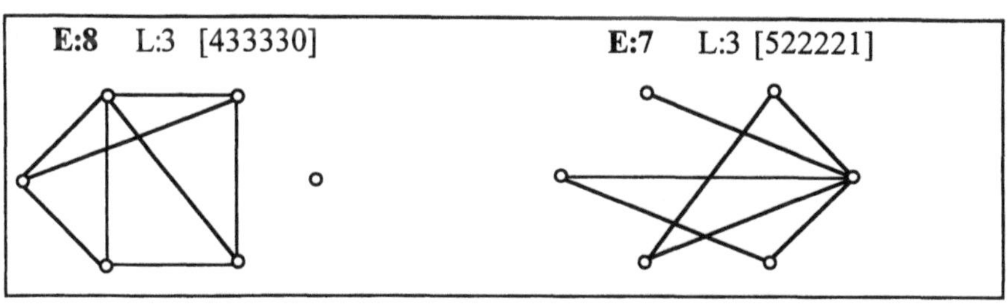
E:8 L:3 [433330]
E:7 L:3 [522221]

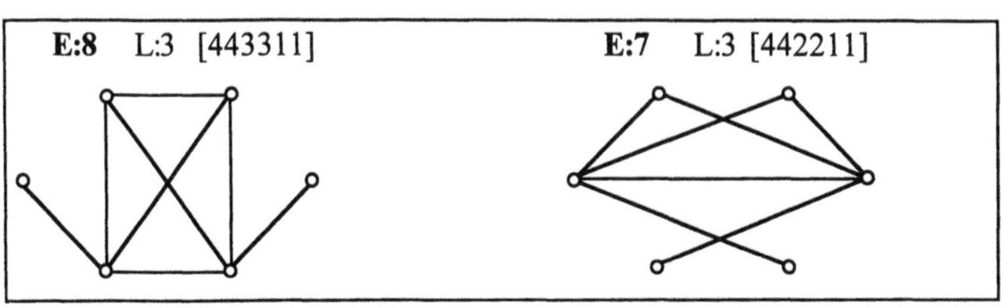
E:8 L:3 [443311]
E:7 L:3 [442211]

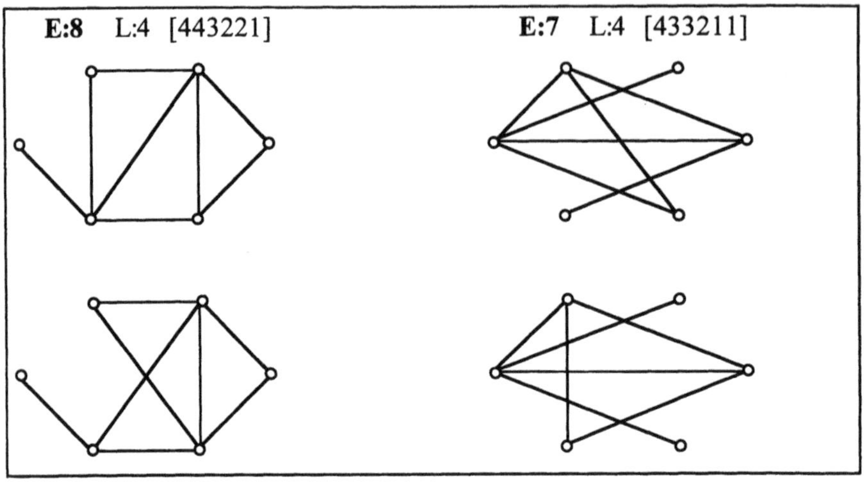
E:8 L:4 [443221]
E:7 L:4 [433211]

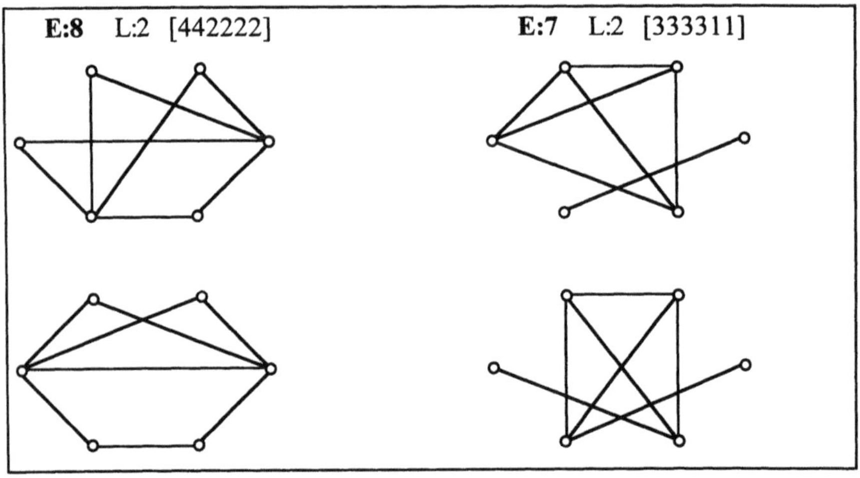
E:8 L:2 [442222]
E:7 L:2 [333311]

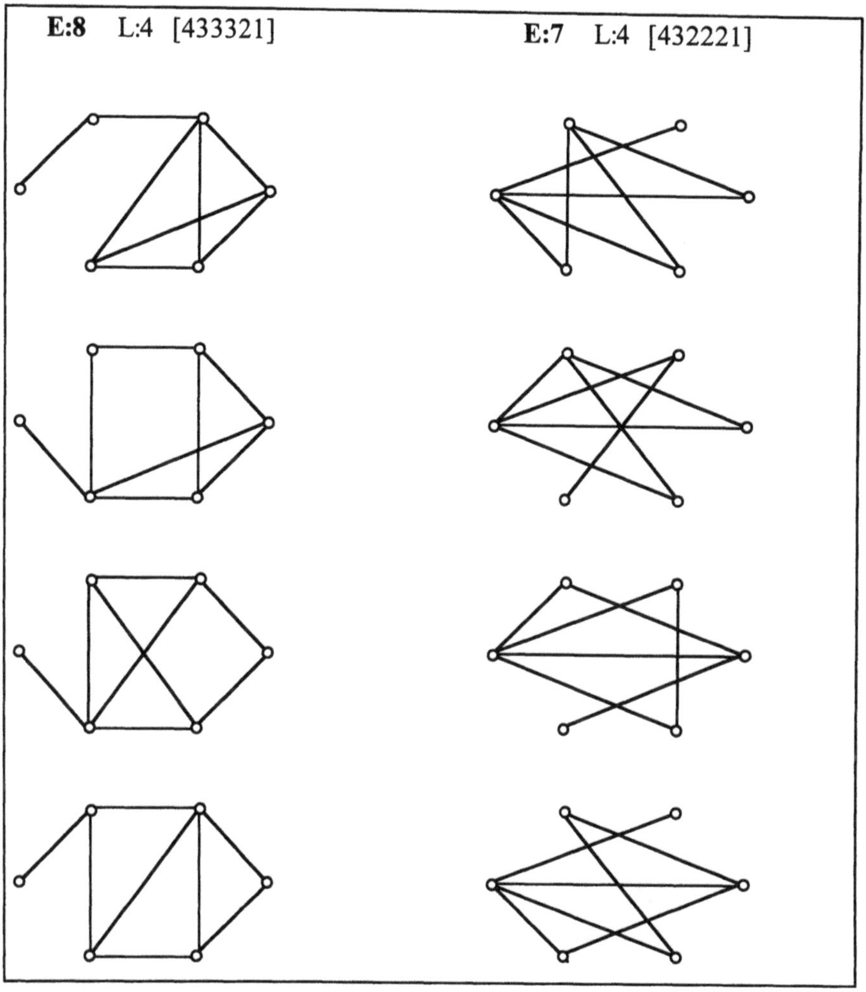
E:8 L:4 [433321]
E:7 L:4 [432221]

E:8 L:3 [433222]

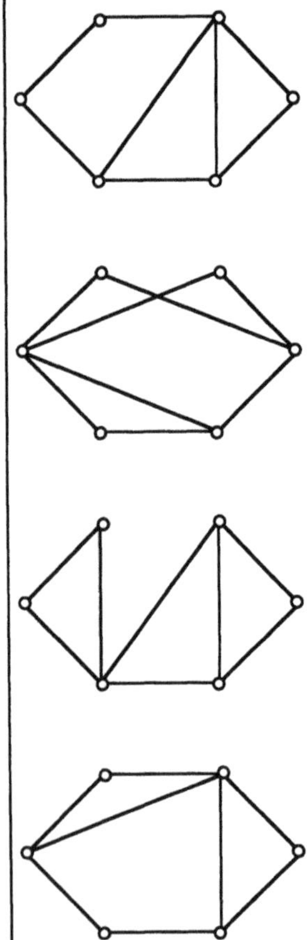

E:7 L:3 [333221]

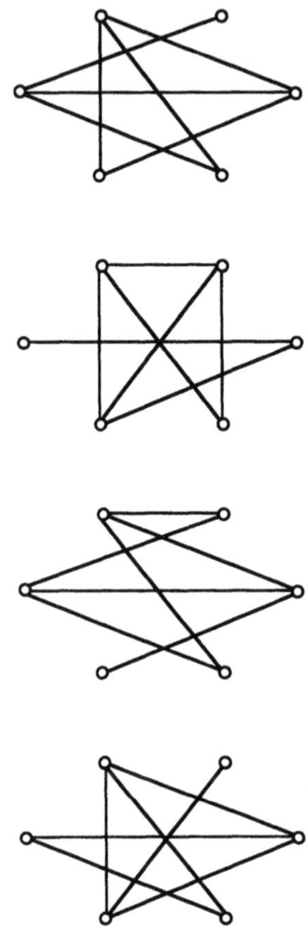

E:8 L:2 [333331]

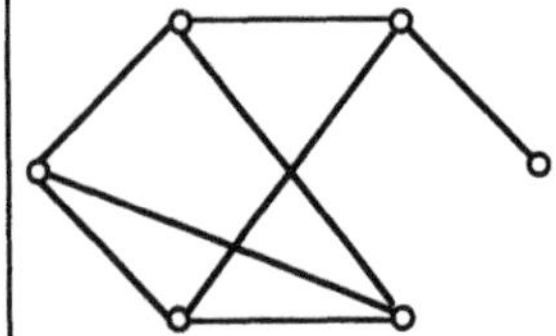

E:7 L:2 [422222]

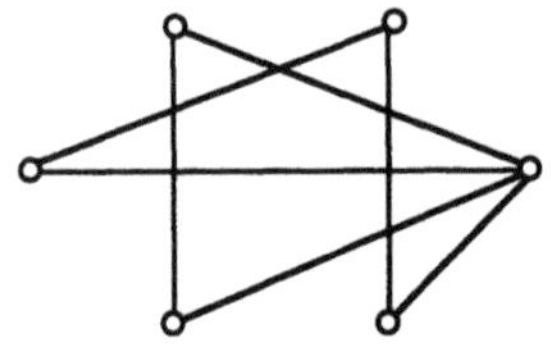

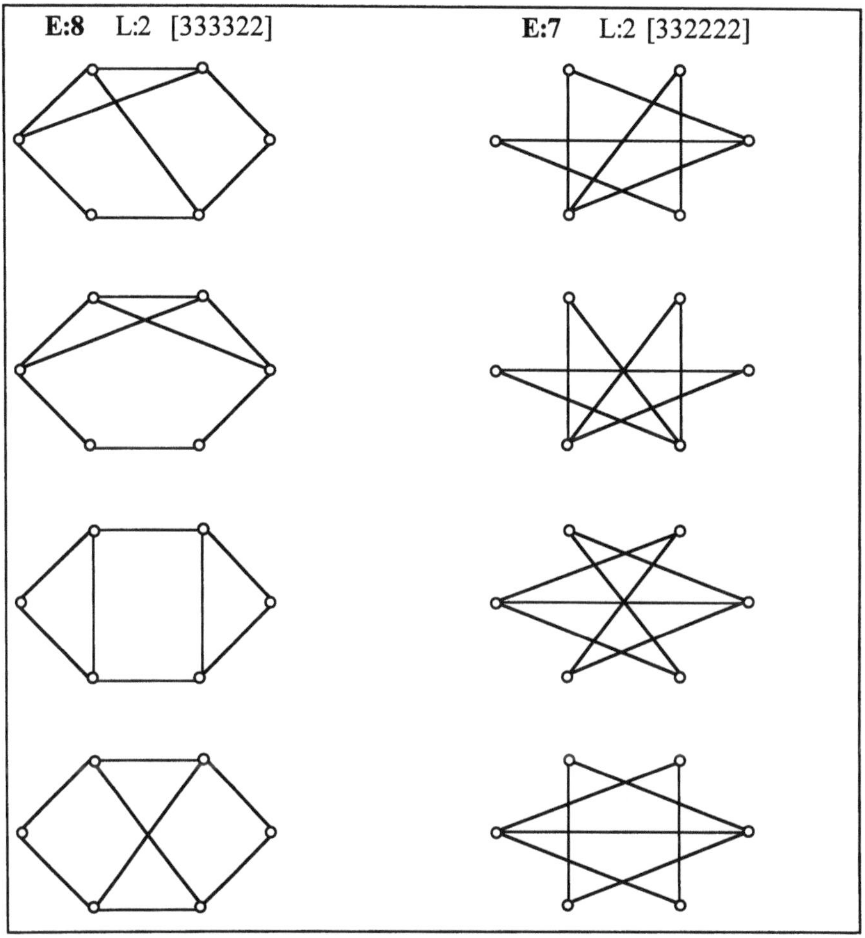
E:8 L:2 [333322]
E:7 L:2 [332222]

Appendix II

12-POSITION DEGREE SEQUENCES FROM FIVE-REGULAR

Each line contains three columns: a degree sequence, the number of hierarchic levels (L:), and the total number of edges required to create the graph (E:).

[555555555555]	L:1	E:30
[555555555553]	L:2	E:29
[555555555551]	L:2	E:28
[555555555544]	L:2	E:29
[555555555542]	L:3	E:28
[555555555533]	L:2	E:28
[555555555531]	L:3	E:27
[555555555522]	L:2	E:27
[555555555511]	L:2	E:26
[555555555443]	L:3	E:28
[555555555441]	L:3	E:27
[555555555432]	L:4	E:27
[555555555421]	L:4	E:26
[555555555333]	L:2	E:27
[555555555331]	L:3	E:26
[555555555322]	L:3	E:26
[555555555311]	L:3	E:25
[555555555221]	L:3	E:25
[555555555111]	L:2	E:24
[555555554444]	L:2	E:28
[555555554442]	L:3	E:27
[555555554433]	L:3	E:27
[555555554431]	L:4	E:26
[555555554422]	L:3	E:26
[555555554411]	L:3	E:25
[555555554332]	L:4	E:26
[555555554321]	L:5	E:25
[555555554222]	L:3	E:25
[555555554211]	L:4	E:24
[555555553333]	L:2	E:26
[555555553331]	L:3	E:25
[555555553322]	L:3	E:25
[555555553311]	L:3	E:24
[555555553221]	L:4	E:24
[555555553111]	L:3	E:23
[555555552222]	L:2	E:24
[555555552211]	L:3	E:23
[555555551111]	L:2	E:22
[555555544443]	L:3	E:27
[555555544441]	L:3	E:26
[555555544432]	L:4	E:26
[555555544421]	L:4	E:25
[555555544333]	L:3	E:26
[555555544331]	L:4	E:25
[555555544322]	L:4	E:25
[555555544311]	L:4	E:24
[555555544221]	L:4	E:24
[555555544111]	L:3	E:23
[555555543332]	L:4	E:25
[555555543321]	L:5	E:24
[555555543222]	L:4	E:24
[555555543211]	L:5	E:23
[555555542221]	L:4	E:23
[555555542111]	L:4	E:22
[555555533333]	L:2	E:25
[555555533331]	L:3	E:24
[555555533322]	L:3	E:24
[555555533311]	L:3	E:23
[555555533221]	L:4	E:23
[555555533111]	L:3	E:22
[555555532222]	L:3	E:23
[555555532211]	L:4	E:22
[555555531111]	L:3	E:21
[555555522221]	L:3	E:22
[555555522111]	L:3	E:21
[555555511111]	L:2	E:20
[555555444444]	L:2	E:27
[555555444442]	L:3	E:26
[555555444433]	L:3	E:26
[555555444431]	L:4	E:25
[555555444422]	L:3	E:25
[555555444411]	L:3	E:24
[555555444332]	L:4	E:25
[555555444321]	L:5	E:24
[555555444222]	L:3	E:24
[555555444211]	L:4	E:23
[555555443333]	L:3	E:25
[555555443331]	L:4	E:24
[555555443322]	L:4	E:24
[555555443311]	L:4	E:23
[555555443221]	L:5	E:23
[555555443111]	L:4	E:22
[555555442222]	L:3	E:23
[555555442211]	L:4	E:22
[555555441111]	L:3	E:21
[555555433332]	L:4	E:24
[555555433321]	L:5	E:23
[555555433222]	L:4	E:23
[555555433211]	L:5	E:22
[555555432221]	L:5	E:22

[555555432111]	L:5	E:21
[555555422222]	L:3	E:22
[555555422211]	L:4	E:21
[555555421111]	L:4	E:20
[555555333333]	L:2	E:24
[555555333331]	L:3	E:23
[555555333322]	L:3	E:23
[555555333311]	L:3	E:22
[555555333221]	L:4	E:22
[555555333111]	L:3	E:21
[555555332222]	L:3	E:22
[555555332211]	L:4	E:21
[555555331111]	L:3	E:20
[555555322221]	L:4	E:21
[555555322111]	L:4	E:20
[555555311111]	L:3	E:19
[555555222222]	L:2	E:21
[555555222211]	L:3	E:20
[555555221111]	L:3	E:19
[555555111111]	L:2	E:18
[555554444443]	L:3	E:26
[555554444441]	L:3	E:25
[555554444432]	L:4	E:25
[555554444421]	L:4	E:24
[555554444333]	L:3	E:25
[555554444331]	L:4	E:24
[555554444322]	L:4	E:24
[555554444311]	L:4	E:23
[555554444221]	L:4	E:23
[555554444111]	L:3	E:22
[555554443332]	L:4	E:24
[555554443321]	L:5	E:23
[555554443222]	L:4	E:23
[555554443211]	L:5	E:22
[555554442221]	L:4	E:22
[555554442111]	L:4	E:21
[555554433333]	L:3	E:24
[555554433331]	L:4	E:23
[555554433322]	L:4	E:23
[555554433311]	L:4	E:22
[555554433221]	L:5	E:22
[555554433111]	L:4	E:21
[555554432222]	L:4	E:22
[555554432211]	L:5	E:21
[555554431111]	L:4	E:20

[555554422221]	L:4	E:21
[555554422111]	L:4	E:20
[555554411111]	L:3	E:19
[555554333332]	L:4	E:23
[555554333321]	L:5	E:22
[555554333222]	L:4	E:22
[555554333211]	L:5	E:21
[555554332221]	L:5	E:21
[555554332111]	L:5	E:20
[555554322222]	L:4	E:21
[555554322211]	L:5	E:20
[555554321111]	L:5	E:19
[555554222221]	L:4	E:20
[555554222111]	L:4	E:19
[555554211111]	L:4	E:18
[555553333333]	L:2	E:23
[555553333331]	L:3	E:22
[555553333322]	L:3	E:22
[555553333311]	L:3	E:21
[555553333221]	L:4	E:21
[555553333111]	L:3	E:20
[555553332222]	L:3	E:21
[555553332211]	L:4	E:20
[555553331111]	L:3	E:19
[555553322221]	L:4	E:20
[555553322111]	L:4	E:19
[555553311111]	L:3	E:18
[555553222222]	L:3	E:20
[555553222211]	L:4	E:19
[555553221111]	L:4	E:18
[555553111111]	L:3	E:17
[555552222221]	L:3	E:19
[555552222111]	L:3	E:18
[555552211111]	L:3	E:17
[555551111111]	L:2	E:16
[555544444444]	L:2	E:26
[555544444442]	L:3	E:25
[555544444433]	L:3	E:25
[555544444431]	L:4	E:24
[555544444422]	L:3	E:24
[555544444411]	L:3	E:23
[555544444332]	L:4	E:24
[555544444321]	L:5	E:23
[555544444222]	L:3	E:23
[555544444211]	L:4	E:22

[555544443333]	L:3	E:24
[555544443331]	L:4	E:23
[555544443322]	L:4	E:23
[555544443311]	L:4	E:22
[555544443221]	L:5	E:22
[555544443111]	L:4	E:21
[555544442222]	L:3	E:22
[555544442211]	L:4	E:21
[555544441111]	L:3	E:20
[555544433332]	L:4	E:23
[555544433321]	L:5	E:22
[555544433222]	L:4	E:22
[555544433211]	L:5	E:21
[555544432221]	L:5	E:21
[555544432111]	L:5	E:20
[555544422222]	L:3	E:21
[555544422211]	L:4	E:20
[555544421111]	L:4	E:19
[555544333333]	L:3	E:23
[555544333331]	L:4	E:22
[555544333322]	L:4	E:22
[555544333311]	L:4	E:21
[555544333221]	L:5	E:21
[555544333111]	L:4	E:20
[555544332222]	L:4	E:21
[555544332211]	L:5	E:20
[555544331111]	L:4	E:19
[555544322221]	L:5	E:20
[555544322111]	L:5	E:19
[555544311111]	L:4	E:18
[555544222222]	L:3	E:20
[555544222211]	L:4	E:19
[555544221111]	L:4	E:18
[555544111111]	L:3	E:17
[555543333332]	L:4	E:22
[555543333321]	L:5	E:21
[555543333222]	L:4	E:21
[555543333211]	L:5	E:20
[555543332221]	L:5	E:20
[555543332111]	L:5	E:19
[555543322222]	L:4	E:20
[555543322211]	L:5	E:19
[555543321111]	L:5	E:18
[555543222221]	L:5	E:19
[555543222111]	L:5	E:18
[555543211111]	L:5	E:17
[555542222222]	L:3	E:19
[555542222211]	L:4	E:18
[555542221111]	L:4	E:17
[555542111111]	L:4	E:16
[555533333333]	L:2	E:22
[555533333331]	L:3	E:21
[555533333322]	L:3	E:21
[555533333311]	L:3	E:20
[555533333221]	L:4	E:20
[555533333111]	L:3	E:19
[555533332222]	L:3	E:20
[555533332211]	L:4	E:19
[555533331111]	L:3	E:18
[555533322221]	L:4	E:19
[555533322111]	L:4	E:18
[555533311111]	L:3	E:17
[555533222222]	L:3	E:19
[555533222211]	L:4	E:18
[555533221111]	L:4	E:17
[555533111111]	L:3	E:16
[555532222221]	L:4	E:18
[555532222111]	L:4	E:17
[555532211111]	L:4	E:16
[555531111111]	L:3	E:15
[555522222222]	L:2	E:18
[555522222211]	L:3	E:17
[555522221111]	L:3	E:16
[555522111111]	L:3	E:15
[555511111111]	L:2	E:14
[555444444443]	L:3	E:25
[555444444441]	L:3	E:24
[555444444432]	L:4	E:24
[555444444421]	L:4	E:23
[555444444333]	L:3	E:24
[555444444331]	L:4	E:23
[555444444322]	L:4	E:23
[555444444311]	L:4	E:22
[555444444221]	L:4	E:22
[555444444111]	L:3	E:21
[555444443332]	L:4	E:23
[555444443321]	L:5	E:22
[555444443222]	L:4	E:22
[555444443211]	L:5	E:21
[555444442221]	L:4	E:21

[5554444421111]	L:4	E:20
[555444433333]	L:3	E:23
[555444433331]	L:4	E:22
[555444433322]	L:4	E:22
[555444433311]	L:4	E:21
[555444433221]	L:5	E:21
[555444433111]	L:4	E:20
[555444432222]	L:4	E:21
[555444432211]	L:5	E:20
[555444431111]	L:4	E:19
[555444422221]	L:4	E:20
[555444422111]	L:4	E:19
[555444411111]	L:3	E:18
[555444333332]	L:4	E:22
[555444333321]	L:5	E:21
[555444333222]	L:4	E:21
[555444333211]	L:5	E:20
[555444332221]	L:5	E:20
[555444332111]	L:5	E:19
[555444322222]	L:4	E:20
[555444322211]	L:5	E:19
[555444321111]	L:5	E:18
[555444222221]	L:4	E:19
[555444222111]	L:4	E:18
[555444211111]	L:4	E:17
[555443333333]	L:3	E:22
[555443333331]	L:4	E:21
[555443333322]	L:4	E:21
[555443333311]	L:4	E:20
[555443333221]	L:5	E:20
[555443333111]	L:4	E:19
[555443332222]	L:4	E:20
[555443332211]	L:5	E:19
[555443331111]	L:4	E:18
[555443322221]	L:5	E:19
[555443322111]	L:5	E:18
[555443311111]	L:4	E:17
[555443222222]	L:4	E:19
[555443222211]	L:5	E:18
[555443221111]	L:5	E:17
[555443111111]	L:4	E:16
[555442222221]	L:4	E:18
[555442222111]	L:4	E:17
[555442211111]	L:4	E:16
[555441111111]	L:3	E:15
[555433333332]	L:4	E:21
[555433333321]	L:5	E:20
[555433333222]	L:4	E:20
[555433333211]	L:5	E:19
[555433332221]	L:5	E:19
[555433332111]	L:5	E:18
[555433322222]	L:4	E:19
[555433322211]	L:5	E:18
[555433321111]	L:5	E:17
[555433222221]	L:5	E:18
[555433222111]	L:5	E:17
[555433211111]	L:5	E:16
[555432222222]	L:4	E:18
[555432222211]	L:5	E:17
[555432221111]	L:5	E:16
[555432111111]	L:5	E:15
[555422222221]	L:4	E:17
[555422222111]	L:4	E:16
[555422211111]	L:4	E:15
[555421111111]	L:4	E:14
[555333333333]	L:2	E:21
[555333333331]	L:3	E:20
[555333333322]	L:3	E:20
[555333333311]	L:3	E:19
[555333333221]	L:4	E:19
[555333333111]	L:3	E:18
[555333332222]	L:3	E:19
[555333332211]	L:4	E:18
[555333331111]	L:3	E:17
[555333322221]	L:4	E:18
[555333322111]	L:4	E:17
[555333311111]	L:3	E:16
[555333222222]	L:3	E:18
[555333222211]	L:4	E:17
[555333221111]	L:4	E:16
[555333111111]	L:3	E:15
[555332222221]	L:4	E:17
[555332222111]	L:4	E:16
[555332211111]	L:4	E:15
[555331111111]	L:3	E:14
[555322222222]	L:3	E:17
[555322222211]	L:4	E:16
[555322221111]	L:4	E:15
[555322111111]	L:4	E:14
[555311111111]	L:3	E:13

[555222222221]	L:3	E:16
[555222222111]	L:3	E:15
[555222211111]	L:3	E:14
[555221111111]	L:3	E:13
[555111111111]	L:2	E:12
[554444444444]	L:2	E:25
[554444444442]	L:3	E:24
[554444444433]	L:3	E:24
[554444444431]	L:4	E:23
[554444444422]	L:3	E:23
[554444444411]	L:3	E:22
[554444444332]	L:4	E:23
[554444444321]	L:5	E:22
[554444444222]	L:3	E:22
[554444444211]	L:4	E:21
[554444443333]	L:3	E:23
[554444443331]	L:4	E:22
[554444443322]	L:4	E:22
[554444443311]	L:4	E:21
[554444443221]	L:5	E:21
[554444443111]	L:4	E:20
[554444442222]	L:3	E:21
[554444442211]	L:4	E:20
[554444441111]	L:3	E:19
[554444433332]	L:4	E:22
[554444433321]	L:5	E:21
[554444433222]	L:4	E:21
[554444433211]	L:5	E:20
[554444432221]	L:5	E:20
[554444432111]	L:5	E:19
[554444422222]	L:3	E:20
[554444422211]	L:4	E:19
[554444421111]	L:4	E:18
[554444333333]	L:3	E:22
[554444333331]	L:4	E:21
[554444333322]	L:4	E:21
[554444333311]	L:4	E:20
[554444333221]	L:5	E:20
[554444333111]	L:4	E:19
[554444332222]	L:4	E:20
[554444332211]	L:5	E:19
[554444331111]	L:4	E:18
[554444322221]	L:5	E:19
[554444322111]	L:5	E:18
[554444311111]	L:4	E:17
[554444222222]	L:3	E:19
[554444222211]	L:4	E:18
[554444221111]	L:4	E:17
[554444111111]	L:3	E:16
[554443333332]	L:4	E:21
[554443333321]	L:5	E:20
[554443333222]	L:4	E:20
[554443333211]	L:5	E:19
[554443332221]	L:5	E:19
[554443332111]	L:5	E:18
[554443322222]	L:4	E:19
[554443322211]	L:5	E:18
[554443321111]	L:5	E:17
[554443222221]	L:5	E:18
[554443222111]	L:5	E:17
[554443211111]	L:5	E:16
[554442222222]	L:3	E:18
[554442222211]	L:4	E:17
[554442221111]	L:4	E:16
[554442111111]	L:4	E:15
[554433333333]	L:3	E:21
[554433333331]	L:4	E:20
[554433333322]	L:4	E:20
[554433333311]	L:4	E:19
[554433333221]	L:5	E:19
[554433333111]	L:4	E:18
[554433332222]	L:4	E:19
[554433332211]	L:5	E:18
[554433331111]	L:4	E:17
[554433322221]	L:5	E:18
[554433322111]	L:5	E:17
[554433311111]	L:4	E:16
[554433222222]	L:4	E:18
[554433222211]	L:5	E:17
[554433221111]	L:5	E:16
[554433111111]	L:4	E:15
[554432222221]	L:5	E:17
[554432222111]	L:5	E:16
[554432211111]	L:5	E:15
[554431111111]	L:4	E:14
[554422222222]	L:3	E:17
[554422222211]	L:4	E:16
[554422221111]	L:4	E:15
[554422111111]	L:4	E:14
[554411111111]	L:3	E:13

[554333333332]	L:4	E:20
[554333333321]	L:5	E:19
[554333333222]	L:4	E:19
[554333333211]	L:5	E:18
[554333332221]	L:5	E:18
[554333332111]	L:5	E:17
[554333322222]	L:4	E:18
[554333322211]	L:5	E:17
[554333321111]	L:5	E:16
[554333222221]	L:5	E:17
[554333222111]	L:5	E:16
[554333211111]	L:5	E:15
[554332222222]	L:4	E:17
[554332222211]	L:5	E:16
[554332221111]	L:5	E:15
[554332111111]	L:5	E:14
[554322222221]	L:5	E:16
[554322222111]	L:5	E:15
[554322211111]	L:5	E:14
[554321111111]	L:5	E:13
[554222222222]	L:3	E:16
[554222222211]	L:4	E:15
[554222221111]	L:4	E:14
[554222111111]	L:4	E:13
[554211111111]	L:4	E:12
[553333333333]	L:2	E:20
[553333333331]	L:3	E:19
[553333333322]	L:3	E:19
[553333333311]	L:3	E:18
[553333333221]	L:4	E:18
[553333333111]	L:3	E:17
[553333332222]	L:3	E:18
[553333332211]	L:4	E:17
[553333331111]	L:3	E:16
[553333322221]	L:4	E:17
[553333322111]	L:4	E:16
[553333311111]	L:3	E:15
[553333222222]	L:3	E:17
[553333222211]	L:4	E:16
[553333221111]	L:4	E:15
[553333111111]	L:3	E:14
[553332222221]	L:4	E:16
[553332222111]	L:4	E:15
[553332211111]	L:4	E:14
[553331111111]	L:3	E:13
[553322222222]	L:3	E:16
[553322222211]	L:4	E:15
[553322221111]	L:4	E:14
[553322111111]	L:4	E:13
[553311111111]	L:3	E:12
[553222222221]	L:4	E:15
[553222222111]	L:4	E:14
[553222211111]	L:4	E:13
[553221111111]	L:4	E:12
[553111111111]	L:3	E:11
[552222222222]	L:2	E:15
[552222222211]	L:3	E:14
[552222221111]	L:3	E:13
[552222111111]	L:3	E:12
[552211111111]	L:3	E:11
[544444444443]	L:3	E:24
[544444444441]	L:3	E:23
[544444444432]	L:4	E:23
[544444444421]	L:4	E:22
[544444444333]	L:3	E:23
[544444444331]	L:4	E:22
[544444444322]	L:4	E:22
[544444444311]	L:4	E:21
[544444444221]	L:4	E:21
[544444444111]	L:3	E:20
[544444443332]	L:4	E:22
[544444443321]	L:5	E:21
[544444443222]	L:4	E:21
[544444443211]	L:5	E:20
[544444442221]	L:4	E:20
[544444442111]	L:4	E:19
[544444433333]	L:3	E:22
[544444433331]	L:4	E:21
[544444433322]	L:4	E:21
[544444433311]	L:4	E:20
[544444433221]	L:5	E:20
[544444433111]	L:4	E:19
[544444432222]	L:4	E:20
[544444432211]	L:5	E:19
[544444431111]	L:4	E:18
[544444422221]	L:4	E:19
[544444422111]	L:4	E:18
[544444411111]	L:3	E:17
[544444333332]	L:4	E:21
[544444333321]	L:5	E:20

[544444333222]	L:4	E:20
[544444333211]	L:5	E:19
[544444332221]	L:5	E:19
[544444332111]	L:5	E:18
[544444322222]	L:4	E:19
[544444322211]	L:5	E:18
[544444321111]	L:5	E:17
[544444222221]	L:4	E:18
[544444222111]	L:4	E:17
[544444211111]	L:4	E:16
[544443333333]	L:3	E:21
[544443333331]	L:4	E:20
[544443333322]	L:4	E:20
[544443333311]	L:4	E:19
[544443333221]	L:5	E:19
[544443333111]	L:4	E:18
[544443332222]	L:4	E:19
[544443332211]	L:5	E:18
[544443331111]	L:4	E:17
[544443322221]	L:5	E:18
[544443322111]	L:5	E:17
[544443311111]	L:4	E:16
[544443222222]	L:4	E:18
[544443222211]	L:5	E:17
[544443221111]	L:5	E:16
[544443111111]	L:4	E:15
[544442222221]	L:4	E:17
[544442222111]	L:4	E:16
[544442211111]	L:4	E:15
[544441111111]	L:3	E:14
[544433333332]	L:4	E:20
[544433333321]	L:5	E:19
[544433333222]	L:4	E:19
[544433333211]	L:5	E:18
[544433332221]	L:5	E:18
[544433332111]	L:5	E:17
[544433322222]	L:4	E:18
[544433322211]	L:5	E:17
[544433321111]	L:5	E:16
[544433222221]	L:5	E:17
[544433222111]	L:5	E:16
[544433211111]	L:5	E:15
[544432222222]	L:4	E:17
[544432222211]	L:5	E:16
[544432221111]	L:5	E:15
[544432111111]	L:5	E:14
[544422222221]	L:4	E:16
[544422222111]	L:4	E:15
[544422211111]	L:4	E:14
[544421111111]	L:4	E:13
[544333333333]	L:3	E:20
[544333333331]	L:4	E:19
[544333333322]	L:4	E:19
[544333333311]	L:4	E:18
[544333333221]	L:5	E:18
[544333333111]	L:4	E:17
[544333332222]	L:4	E:18
[544333332211]	L:5	E:17
[544333331111]	L:4	E:16
[544333322221]	L:5	E:17
[544333322111]	L:5	E:16
[544333311111]	L:4	E:15
[544333222222]	L:4	E:17
[544333222211]	L:5	E:16
[544333221111]	L:5	E:15
[544333111111]	L:4	E:14
[544332222221]	L:5	E:16
[544332222111]	L:5	E:15
[544332211111]	L:5	E:14
[544331111111]	L:4	E:13
[544322222222]	L:4	E:16
[544322222211]	L:5	E:15
[544322221111]	L:5	E:14
[544322111111]	L:5	E:13
[544311111111]	L:4	E:12
[544222222221]	L:4	E:15
[544222222111]	L:4	E:14
[544222211111]	L:4	E:13
[544221111111]	L:4	E:12
[544111111111]	L:3	E:11
[543333333332]	L:4	E:19
[543333333321]	L:5	E:18
[543333333222]	L:4	E:18
[543333333211]	L:5	E:17
[543333332221]	L:5	E:17
[543333332111]	L:5	E:16
[543333322222]	L:4	E:17
[543333322211]	L:5	E:16
[543333321111]	L:5	E:15
[543333222221]	L:5	E:16

[543333222111]	L:5	E:15
[543333211111]	L:5	E:14
[543332222222]	L:4	E:16
[543332222211]	L:5	E:15
[543332221111]	L:5	E:14
[543332111111]	L:5	E:13
[543322222221]	L:5	E:15
[543322222111]	L:5	E:14
[543322211111]	L:5	E:13
[543321111111]	L:5	E:12
[543222222222]	L:4	E:15
[543222222211]	L:5	E:14
[543222221111]	L:5	E:13
[543222111111]	L:5	E:12
[543211111111]	L:5	E:11
[542222222221]	L:4	E:14
[542222222111]	L:4	E:13
[542222211111]	L:4	E:12
[542221111111]	L:4	E:11
[533333333333]	L:2	E:19
[533333333331]	L:3	E:18
[533333333322]	L:3	E:18
[533333333311]	L:3	E:17
[533333333221]	L:4	E:17
[533333333111]	L:3	E:16
[533333332222]	L:3	E:17
[533333332211]	L:4	E:16
[533333331111]	L:3	E:15
[533333322221]	L:4	E:16
[533333322111]	L:4	E:15
[533333311111]	L:3	E:14
[533333222222]	L:3	E:16
[533333222211]	L:4	E:15
[533333221111]	L:4	E:14
[533333111111]	L:3	E:13
[533332222221]	L:4	E:15
[533332222111]	L:4	E:14
[533332211111]	L:4	E:13
[533331111111]	L:3	E:12
[533322222222]	L:3	E:15
[533322222211]	L:4	E:14
[533322221111]	L:4	E:13
[533322111111]	L:4	E:12
[533311111111]	L:3	E:11
[533222222221]	L:4	E:14
[533222222111]	L:4	E:13
[533222211111]	L:4	E:12
[533221111111]	L:4	E:11
[532222222222]	L:3	E:14
[532222222211]	L:4	E:13
[532222221111]	L:4	E:12
[532222111111]	L:4	E:11
[522222222221]	L:3	E:13
[522222222111]	L:3	E:12
[522222211111]	L:3	E:11
[444444444444]	L:1	E:24
[444444444442]	L:2	E:23
[444444444433]	L:2	E:23
[444444444431]	L:3	E:22
[444444444422]	L:2	E:22
[444444444411]	L:2	E:21
[444444444332]	L:3	E:22
[444444444321]	L:4	E:21
[444444444222]	L:2	E:21
[444444444211]	L:3	E:20
[444444443333]	L:2	E:22
[444444443331]	L:3	E:21
[444444443322]	L:3	E:21
[444444443311]	L:3	E:20
[444444443221]	L:4	E:20
[444444443111]	L:3	E:19
[444444442222]	L:2	E:20
[444444442211]	L:3	E:19
[444444441111]	L:2	E:18
[444444433332]	L:3	E:21
[444444433321]	L:4	E:20
[444444433222]	L:3	E:20
[444444433211]	L:4	E:19
[444444432221]	L:4	E:19
[444444432111]	L:4	E:18
[444444422222]	L:2	E:19
[444444422211]	L:3	E:18
[444444421111]	L:3	E:17
[444444333333]	L:2	E:21
[444444333331]	L:3	E:20
[444444333322]	L:3	E:20
[444444333311]	L:3	E:19
[444444333221]	L:4	E:19
[444444333111]	L:3	E:18
[444444332222]	L:3	E:19

[444444332211]	L:4	E:18
[444444331111]	L:3	E:17
[444444322221]	L:4	E:18
[444444322111]	L:4	E:17
[444444311111]	L:3	E:16
[444444222222]	L:2	E:18
[444444222211]	L:3	E:17
[444444221111]	L:3	E:16
[444444111111]	L:2	E:15
[444443333332]	L:3	E:20
[444443333321]	L:4	E:19
[444443333222]	L:3	E:19
[444443333211]	L:4	E:18
[444443332221]	L:4	E:18
[444443332111]	L:4	E:17
[444443322222]	L:3	E:18
[444443322211]	L:4	E:17
[444443321111]	L:4	E:16
[444443222221]	L:4	E:17
[444443222111]	L:4	E:16
[444443211111]	L:4	E:15
[444442222222]	L:2	E:17
[444442222211]	L:3	E:16
[444442221111]	L:3	E:15
[444442111111]	L:3	E:14
[444433333333]	L:2	E:20
[444433333331]	L:3	E:19
[444433333322]	L:3	E:19
[444433333311]	L:3	E:18
[444433333221]	L:4	E:18
[444433333111]	L:3	E:17
[444433332222]	L:3	E:18
[444433332211]	L:4	E:17
[444433331111]	L:3	E:16
[444433322221]	L:4	E:17
[444433322111]	L:4	E:16
[444433311111]	L:3	E:15
[444433222222]	L:3	E:17
[444433222211]	L:4	E:16
[444433221111]	L:4	E:15
[444433111111]	L:3	E:14
[444432222221]	L:4	E:16
[444432222111]	L:4	E:15
[444432211111]	L:4	E:14
[444431111111]	L:3	E:13
[444422222222]	L:2	E:16
[444422222211]	L:3	E:15
[444422221111]	L:3	E:14
[444422111111]	L:3	E:13
[444411111111]	L:2	E:12
[444333333332]	L:3	E:19
[444333333321]	L:4	E:18
[444333333222]	L:3	E:18
[444333333211]	L:4	E:17
[444333332221]	L:4	E:17
[444333332111]	L:4	E:16
[444333322222]	L:3	E:17
[444333322211]	L:4	E:16
[444333321111]	L:4	E:15
[444333222221]	L:4	E:16
[444333222111]	L:4	E:15
[444333211111]	L:4	E:14
[444332222222]	L:3	E:16
[444332222211]	L:4	E:15
[444332221111]	L:4	E:14
[444332111111]	L:4	E:13
[444322222221]	L:4	E:15
[444322222111]	L:4	E:14
[444322211111]	L:4	E:13
[444321111111]	L:4	E:12
[444222222222]	L:2	E:15
[444222222211]	L:3	E:14
[444222221111]	L:3	E:13
[444222111111]	L:3	E:12
[444211111111]	L:3	E:11
[443333333333]	L:2	E:19
[443333333331]	L:3	E:18
[443333333322]	L:3	E:18
[443333333311]	L:3	E:17
[443333333221]	L:4	E:17
[443333333111]	L:3	E:16
[443333332222]	L:3	E:17
[443333332211]	L:4	E:16
[443333331111]	L:3	E:15
[443333322221]	L:4	E:16
[443333322111]	L:4	E:15
[443333311111]	L:3	E:14
[443333222222]	L:3	E:16
[443333222211]	L:4	E:15
[443333221111]	L:4	E:14

[443333111111]	L:3	E:13
[443332222221]	L:4	E:15
[443332222111]	L:4	E:14
[443332211111]	L:4	E:13
[443331111111]	L:3	E:12
[443322222222]	L:3	E:15
[443322222211]	L:4	E:14
[443322221111]	L:4	E:13
[443322111111]	L:4	E:12
[443311111111]	L:3	E:11
[443222222221]	L:4	E:14
[443222222111]	L:4	E:13
[443222211111]	L:4	E:12
[443221111111]	L:4	E:11
[442222222222]	L:2	E:14
[442222222211]	L:3	E:13
[442222221111]	L:3	E:12
[442222111111]	L:3	E:11
[433333333332]	L:3	E:18
[433333333321]	L:4	E:17
[433333333222]	L:3	E:17
[433333333211]	L:4	E:16
[433333332221]	L:4	E:16
[433333332111]	L:4	E:15
[433333322222]	L:3	E:16
[433333322211]	L:4	E:15
[433333321111]	L:4	E:14
[433333222221]	L:4	E:15
[433333222111]	L:4	E:14
[433333211111]	L:4	E:13
[433332222222]	L:3	E:15
[433332222211]	L:4	E:14
[433332221111]	L:4	E:13
[433332111111]	L:4	E:12
[433322222221]	L:4	E:14
[433322222111]	L:4	E:13
[433322211111]	L:4	E:12
[433321111111]	L:4	E:11
[433222222222]	L:3	E:14
[433222222211]	L:4	E:13
[433222221111]	L:4	E:12
[433222111111]	L:4	E:11
[432222222221]	L:4	E:13
[432222222111]	L:4	E:12
[432222211111]	L:4	E:11
[422222222222]	L:2	E:13
[422222222211]	L:3	E:12
[422222221111]	L:3	E:11
[333333333333]	L:1	E:18
[333333333331]	L:2	E:17
[333333333322]	L:2	E:17
[333333333311]	L:2	E:16
[333333333221]	L:3	E:16
[333333333111]	L:2	E:15
[333333332222]	L:2	E:16
[333333332211]	L:3	E:15
[333333331111]	L:2	E:14
[333333322221]	L:3	E:15
[333333322111]	L:3	E:14
[333333311111]	L:2	E:13
[333333222222]	L:2	E:15
[333333222211]	L:3	E:14
[333333221111]	L:3	E:13
[333333111111]	L:2	E:12
[333332222221]	L:3	E:14
[333332222111]	L:3	E:13
[333332211111]	L:3	E:12
[333331111111]	L:2	E:11
[333322222222]	L:2	E:14
[333322222211]	L:3	E:13
[333322221111]	L:3	E:12
[333322111111]	L:3	E:11
[333222222221]	L:3	E:13
[333222222111]	L:3	E:12
[333222211111]	L:3	E:11
[332222222222]	L:2	E:13
[332222222211]	L:3	E:12
[332222221111]	L:3	E:11
[322222222221]	L:3	E:12
[322222222111]	L:3	E:11
[222222222222]	L:1	E:12
[222222222211]	L:2	E:11

Appendix III

SELECTED 12-POSITION DEGREE SEQUENCES BY EDGES REQUIRED

Each line contains three columns: a degree sequence, the number of hierarchic levels (L:), and the total number of edges required to create the graph (E:).

Factor-11 Sequences

[B11111111111]	L:2	E:11
[A21111111111]	L:3	E:11
[931111111111]	L:3	E:11
[922111111111]	L:3	E:11
[841111111111]	L:2	E:11
[832111111111]	L:3	E:11
[822211111111]	L:2	E:11
[751111111111]	L:3	E:11
[733111111111]	L:3	E:11
[722221111111]	L:3	E:11
[742111111111]	L:4	E:11
[732211111111]	L:4	E:11
[661111111111]	L:2	E:11
[622222111111]	L:3	E:11
[652111111111]	L:4	E:11
[643111111111]	L:4	E:11
[642211111111]	L:4	E:11
[633211111111]	L:4	E:11
[632221111111]	L:4	E:11
[553111111111]	L:3	E:11
[552211111111]	L:3	E:11
[544111111111]	L:3	E:11
[543211111111]	L:5	E:11
[542221111111]	L:4	E:11
[533311111111]	L:3	E:11
[533221111111]	L:4	E:11
[532222111111]	L:4	E:11
[522222211111]	L:3	E:11
[444211111111]	L:3	E:11
[443311111111]	L:3	E:11
[443221111111]	L:4	E:11
[442222111111]	L:3	E:11
[433321111111]	L:4	E:11
[433222111111]	L:4	E:11
[432222211111]	L:4	E:11
[422222221111]	L:3	E:11
[333331111111]	L:2	E:11
[333322111111]	L:3	E:11
[333222211111]	L:3	E:11
[332222221111]	L:3	E:11
[322222222111]	L:3	E:11
[222222222211]	L:2	E:11

Factor-12 Sequences

[B22111111111]	L:3	E:12
[A32111111111]	L:4	E:12
[933111111111]	L:3	E:12
[922221111111]	L:3	E:12
[942111111111]	L:4	E:12
[932211111111]	L:4	E:12
[852111111111]	L:3	E:12
[843111111111]	L:3	E:12
[842211111111]	L:3	E:12
[833211111111]	L:3	E:12
[832221111111]	L:3	E:12
[822222111111]	L:2	E:12
[744111111111]	L:3	E:12
[733311111111]	L:3	E:12
[722222211111]	L:3	E:12
[762111111111]	L:4	E:12
[753111111111]	L:4	E:12
[752211111111]	L:4	E:12
[742221111111]	L:4	E:12
[733221111111]	L:4	E:12
[732222111111]	L:4	E:12
[743211111111]	L:5	E:12
[663111111111]	L:3	E:12
[662211111111]	L:3	E:12
[622222221111]	L:3	E:12
[654111111111]	L:4	E:12
[652221111111]	L:4	E:12
[644211111111]	L:4	E:12
[643311111111]	L:4	E:12
[642222111111]	L:4	E:12
[633321111111]	L:4	E:12
[633222111111]	L:4	E:12
[632222211111]	L:4	E:12
[653211111111]	L:5	E:12
[643221111111]	L:5	E:12
[555111111111]	L:2	E:12
[553311111111]	L:3	E:12
[552222111111]	L:3	E:12

[533331111111]	L:3	E:12
[522222222111]	L:3	E:12
[554211111111]	L:4	E:12
[553221111111]	L:4	E:12
[544311111111]	L:4	E:12
[544221111111]	L:4	E:12
[542222211111]	L:4	E:12
[533322111111]	L:4	E:12
[533222211111]	L:4	E:12
[532222221111]	L:4	E:12
[543321111111]	L:5	E:12
[543222111111]	L:5	E:12
[444411111111]	L:2	E:12
[444222111111]	L:3	E:12
[443331111111]	L:3	E:12
[442222221111]	L:3	E:12
[422222222211]	L:3	E:12
[444321111111]	L:4	E:12
[443322111111]	L:4	E:12
[443222211111]	L:4	E:12
[433332111111]	L:4	E:12
[433322211111]	L:4	E:12
[433222221111]	L:4	E:12
[432222222111]	L:4	E:12
[333333111111]	L:2	E:12
[333332211111]	L:3	E:12
[333322221111]	L:3	E:12
[333222222111]	L:3	E:12
[332222222211]	L:3	E:12
[322222222221]	L:3	E:12
[222222222222]	L:1	E:12

Factor-13 Sequences

[B22221111111]	L:3	E:13
[B32211111111]	L:4	E:13
[A22222111111]	L:3	E:13
[A42211111111]	L:4	E:13
[A33211111111]	L:4	E:13
[A32221111111]	L:4	E:13
[933311111111]	L:3	E:13
[922222211111]	L:3	E:13
[952211111111]	L:4	E:13
[942221111111]	L:4	E:13
[933221111111]	L:4	E:13
[932222111111]	L:4	E:13
[943211111111]	L:5	E:13
[862211111111]	L:3	E:13
[853211111111]	L:4	E:13
[852221111111]	L:3	E:13
[844211111111]	L:3	E:13
[843311111111]	L:3	E:13
[843221111111]	L:4	E:13
[842222111111]	L:3	E:13
[833321111111]	L:3	E:13
[833222111111]	L:3	E:13
[832222211111]	L:3	E:13
[822222221111]	L:2	E:13
[772211111111]	L:3	E:13
[733331111111]	L:3	E:13
[722222222111]	L:3	E:13
[762221111111]	L:4	E:13
[753311111111]	L:4	E:13
[752222111111]	L:4	E:13
[744311111111]	L:4	E:13
[744221111111]	L:4	E:13
[742222211111]	L:4	E:13
[733322111111]	L:4	E:13
[733222211111]	L:4	E:13
[732222221111]	L:4	E:13
[763211111111]	L:5	E:13
[754211111111]	L:5	E:13
[753221111111]	L:5	E:13
[743321111111]	L:5	E:13
[743222111111]	L:5	E:13
[663311111111]	L:3	E:13
[662222111111]	L:3	E:13
[644411111111]	L:3	E:13
[622222222211]	L:3	E:13
[664211111111]	L:4	E:13
[663221111111]	L:4	E:13
[655211111111]	L:4	E:13
[652222211111]	L:4	E:13
[644222111111]	L:4	E:13
[643331111111]	L:4	E:13
[642222221111]	L:4	E:13

[633332111111]	L:4	E:13
[633322211111]	L:4	E:13
[633222221111]	L:4	E:13
[632222222111]	L:4	E:13
[654311111111]	L:5	E:13
[654221111111]	L:5	E:13
[653321111111]	L:5	E:13
[653222111111]	L:5	E:13
[644321111111]	L:5	E:13
[643322111111]	L:5	E:13
[643222211111]	L:5	E:13
[555311111111]	L:3	E:13
[555221111111]	L:3	E:13
[554411111111]	L:3	E:13
[553331111111]	L:3	E:13
[552222221111]	L:3	E:13
[533333111111]	L:3	E:13
[522222222221]	L:3	E:13
[554222111111]	L:4	E:13
[553322111111]	L:4	E:13
[553222211111]	L:4	E:13
[544421111111]	L:4	E:13
[544331111111]	L:4	E:13
[544222211111]	L:4	E:13
[542222222111]	L:4	E:13
[533332211111]	L:4	E:13
[533322221111]	L:4	E:13
[533222222111]	L:4	E:13
[532222222211]	L:4	E:13
[554321111111]	L:5	E:13
[544322111111]	L:5	E:13
[543332111111]	L:5	E:13
[543322211111]	L:5	E:13
[543222221111]	L:5	E:13
[422222222222]	L:2	E:13
[444431111111]	L:3	E:13
[444422111111]	L:3	E:13
[444222221111]	L:3	E:13
[443333111111]	L:3	E:13
[442222222211]	L:3	E:13
[444332111111]	L:4	E:13
[444322211111]	L:4	E:13
[443332211111]	L:4	E:13
[443322221111]	L:4	E:13
[443222222111]	L:4	E:13
[433333211111]	L:4	E:13
[433332221111]	L:4	E:13
[433322222111]	L:4	E:13
[433222222211]	L:4	E:13
[432222222221]	L:4	E:13
[333333311111]	L:2	E:13
[332222222222]	L:2	E:13
[333333221111]	L:3	E:13
[333332222111]	L:3	E:13
[333322222211]	L:3	E:13
[333222222221]	L:3	E:13

Factor-14 Sequences

[B33311111111]	L:3	E:14
[B22222211111]	L:3	E:14
[B42221111111]	L:4	E:14
[B33221111111]	L:4	E:14
[B32222111111]	L:4	E:14
[A22222221111]	L:3	E:14
[A52221111111]	L:4	E:14
[A43311111111]	L:4	E:14
[A42222111111]	L:4	E:14
[A33321111111]	L:4	E:14
[A33222111111]	L:4	E:14
[A32222211111]	L:4	E:14
[A43221111111]	L:5	E:14
[933331111111]	L:3	E:14
[922222222111]	L:3	E:14
[962221111111]	L:4	E:14
[953311111111]	L:4	E:14
[952222111111]	L:4	E:14
[944311111111]	L:4	E:14
[944221111111]	L:4	E:14
[943222111111]	L:4	E:14
[942222211111]	L:4	E:14
[933322111111]	L:4	E:14
[933222211111]	L:4	E:14
[932222221111]	L:4	E:14
[953221111111]	L:5	E:14
[943321111111]	L:5	E:14

[872221111111]	L:3	E:14
[863311111111]	L:3	E:14
[863221111111]	L:4	E:14
[862222111111]	L:3	E:14
[854311111111]	L:4	E:14
[854221111111]	L:4	E:14
[853321111111]	L:4	E:14
[853222111111]	L:4	E:14
[852222211111]	L:3	E:14
[844411111111]	L:2	E:14
[844321111111]	L:4	E:14
[844222111111]	L:3	E:14
[843331111111]	L:3	E:14
[843322111111]	L:4	E:14
[843222211111]	L:4	E:14
[842222221111]	L:3	E:14
[833332111111]	L:3	E:14
[833322211111]	L:3	E:14
[833222221111]	L:3	E:14
[832222222111]	L:3	E:14
[822222222211]	L:2	E:14
[773311111111]	L:3	E:14
[772222111111]	L:3	E:14
[733333111111]	L:3	E:14
[722222222221]	L:3	E:14
[773221111111]	L:4	E:14
[762222211111]	L:4	E:14
[755311111111]	L:4	E:14
[755221111111]	L:4	E:14
[754411111111]	L:4	E:14
[753331111111]	L:4	E:14
[752222221111]	L:4	E:14
[744421111111]	L:4	E:14
[744331111111]	L:4	E:14
[744222211111]	L:4	E:14
[742222222111]	L:4	E:14
[733332211111]	L:4	E:14
[733322221111]	L:4	E:14
[733222222111]	L:4	E:14
[732222222211]	L:4	E:14
[764311111111]	L:5	E:14
[764221111111]	L:5	E:14
[763321111111]	L:5	E:14
[763222111111]	L:5	E:14
[754222111111]	L:5	E:14
[753322111111]	L:5	E:14
[753222211111]	L:5	E:14
[744322111111]	L:5	E:14
[743332111111]	L:5	E:14
[743322211111]	L:5	E:14
[743222221111]	L:5	E:14
[754321111111]	L:6	E:14
[622222222222]	L:2	E:14
[664411111111]	L:3	E:14
[663331111111]	L:3	E:14
[662222221111]	L:3	E:14
[665311111111]	L:4	E:14
[665221111111]	L:4	E:14
[664222111111]	L:4	E:14
[663322111111]	L:4	E:14
[663222211111]	L:4	E:14
[655411111111]	L:4	E:14
[655222111111]	L:4	E:14
[652222222111]	L:4	E:14
[644431111111]	L:4	E:14
[644422111111]	L:4	E:14
[644222221111]	L:4	E:14
[643333111111]	L:4	E:14
[642222222211]	L:4	E:14
[633333211111]	L:4	E:14
[633332221111]	L:4	E:14
[633322222111]	L:4	E:14
[633222222211]	L:4	E:14
[632222222221]	L:4	E:14
[664321111111]	L:5	E:14
[655321111111]	L:5	E:14
[654421111111]	L:5	E:14
[654331111111]	L:5	E:14
[654222211111]	L:5	E:14
[653332111111]	L:5	E:14
[653322211111]	L:5	E:14
[653222221111]	L:5	E:14
[644332111111]	L:5	E:14
[644322211111]	L:5	E:14
[643332211111]	L:5	E:14
[643322221111]	L:5	E:14

[643222222111]	L:5	E:14
[654322111111]	L:6	E:14
[555511111111]	L:2	E:14
[555331111111]	L:3	E:14
[555222211111]	L:3	E:14
[553333111111]	L:3	E:14
[552222222211]	L:3	E:14
[544441111111]	L:3	E:14
[533333311111]	L:3	E:14
[532222222222]	L:3	E:14
[555421111111]	L:4	E:14
[555322111111]	L:4	E:14
[554431111111]	L:4	E:14
[554422111111]	L:4	E:14
[554222221111]	L:4	E:14
[553332211111]	L:4	E:14
[553322221111]	L:4	E:14
[553222222111]	L:4	E:14
[544422211111]	L:4	E:14
[544333111111]	L:4	E:14
[544222222111]	L:4	E:14
[542222222221]	L:4	E:14
[533333221111]	L:4	E:14
[533332222111]	L:4	E:14
[533322222211]	L:4	E:14
[533222222221]	L:4	E:14
[554332111111]	L:5	E:14
[554322211111]	L:5	E:14
[544432111111]	L:5	E:14
[544332211111]	L:5	E:14
[544322221111]	L:5	E:14
[543333211111]	L:5	E:14
[543332221111]	L:5	E:14
[543322222111]	L:5	E:14
[543222222211]	L:5	E:14
[442222222222]	L:2	E:14
[444442111111]	L:3	E:14
[444433111111]	L:3	E:14
[444422221111]	L:3	E:14
[444222222211]	L:3	E:14
[443333311111]	L:3	E:14
[433222222222]	L:3	E:14
[444432211111]	L:4	E:14
[444333211111]	L:4	E:14
[444332221111]	L:4	E:14
[444322222111]	L:4	E:14
[443333221111]	L:4	E:14
[443332222111]	L:4	E:14
[443322222211]	L:4	E:14
[443222222221]	L:4	E:14
[433333321111]	L:4	E:14
[433333222111]	L:4	E:14
[433332222211]	L:4	E:14
[433322222221]	L:4	E:14
[333333331111]	L:2	E:14
[333322222222]	L:2	E:14
[333333322111]	L:3	E:14
[333333222211]	L:3	E:14
[333332222221]	L:3	E:14

Factor-15 Sequences

[B33331111111]	L:3	E:15
[B22222222111]	L:3	E:15
[B52222111111]	L:4	E:15
[B42222211111]	L:4	E:15
[B33322111111]	L:4	E:15
[B33222211111]	L:4	E:15
[B32222221111]	L:4	E:15
[B43321111111]	L:5	E:15
[B43222111111]	L:5	E:15
[A22222222211]	L:3	E:15
[A62222111111]	L:4	E:15
[A52222211111]	L:4	E:15
[A44222111111]	L:4	E:15
[A43331111111]	L:4	E:15
[A43331111111]	L:4	E:15
[A42222221111]	L:4	E:15
[A33332111111]	L:4	E:15
[A33322211111]	L:4	E:15
[A33222221111]	L:4	E:15
[A32222222111]	L:4	E:15
[A53321111111]	L:5	E:15
[A53222111111]	L:5	E:15
[A44321111111]	L:5	E:15
[A43322111111]	L:5	E:15

[A43222211111]	L:5	E:15
[933333111111]	L:3	E:15
[922222222221]	L:3	E:15
[972222111111]	L:4	E:15
[962222211111]	L:4	E:15
[953331111111]	L:4	E:15
[952222221111]	L:4	E:15
[944421111111]	L:4	E:15
[944331111111]	L:4	E:15
[944222211111]	L:4	E:15
[942222222111]	L:4	E:15
[933332211111]	L:4	E:15
[933322221111]	L:4	E:15
[933222222111]	L:4	E:15
[932222222211]	L:4	E:15
[963321111111]	L:5	E:15
[963222111111]	L:5	E:15
[954222111111]	L:5	E:15
[953322111111]	L:5	E:15
[953222211111]	L:5	E:15
[944322111111]	L:5	E:15
[943332111111]	L:5	E:15
[943322211111]	L:5	E:15
[943222221111]	L:5	E:15
[954321111111]	L:6	E:15
[882222111111]	L:2	E:15
[873321111111]	L:4	E:15
[873222111111]	L:4	E:15
[872222211111]	L:3	E:15
[864321111111]	L:5	E:15
[864222111111]	L:4	E:15
[863331111111]	L:3	E:15
[863322111111]	L:4	E:15
[863222211111]	L:4	E:15
[862222221111]	L:3	E:15
[855321111111]	L:4	E:15
[855222111111]	L:3	E:15
[854421111111]	L:4	E:15
[854331111111]	L:4	E:15
[854322111111]	L:5	E:15
[854222211111]	L:4	E:15
[853332111111]	L:4	E:15
[853322211111]	L:4	E:15
[853222221111]	L:4	E:15
[852222222111]	L:3	E:15
[844431111111]	L:3	E:15
[844422111111]	L:3	E:15
[844332111111]	L:4	E:15
[844322211111]	L:4	E:15
[844222221111]	L:3	E:15
[843333111111]	L:3	E:15
[843332211111]	L:4	E:15
[843322221111]	L:4	E:15
[843222222111]	L:4	E:15
[842222222211]	L:3	E:15
[833333211111]	L:3	E:15
[833332221111]	L:3	E:15
[833322222111]	L:3	E:15
[833222222211]	L:3	E:15
[832222222221]	L:4	E:15
[822222222222]	L:2	E:15
[773331111111]	L:3	E:15
[772222221111]	L:3	E:15
[744441111111]	L:3	E:15
[733333311111]	L:3	E:15
[732222222222]	L:3	E:15
[774222111111]	L:4	E:15
[773322111111]	L:4	E:15
[773222211111]	L:4	E:15
[762222222111]	L:4	E:15
[755331111111]	L:4	E:15
[755222211111]	L:4	E:15
[753333111111]	L:4	E:15
[752222222211]	L:4	E:15
[744422211111]	L:4	E:15
[744333111111]	L:4	E:15
[744222222111]	L:4	E:15
[742222222221]	L:4	E:15
[733333221111]	L:4	E:15
[733332222111]	L:4	E:15
[733322222211]	L:4	E:15
[733222222221]	L:4	E:15
[774321111111]	L:5	E:15
[765222111111]	L:5	E:15
[764421111111]	L:5	E:15
[764331111111]	L:5	E:15

[764222211111]	L:5	E:15
[763332111111]	L:5	E:15
[763322211111]	L:5	E:15
[763222221111]	L:5	E:15
[755421111111]	L:5	E:15
[755322111111]	L:5	E:15
[754431111111]	L:5	E:15
[754422111111]	L:5	E:15
[754222221111]	L:5	E:15
[753332211111]	L:5	E:15
[753322221111]	L:5	E:15
[753222222111]	L:5	E:15
[744432111111]	L:5	E:15
[744332211111]	L:5	E:15
[744322221111]	L:5	E:15
[743333211111]	L:5	E:15
[743332221111]	L:5	E:15
[743322222111]	L:5	E:15
[743222222211]	L:5	E:15
[765321111111]	L:6	E:15
[764322111111]	L:6	E:15
[754332111111]	L:6	E:15
[754322211111]	L:6	E:15
[666222111111]	L:3	E:15
[663333111111]	L:3	E:15
[662222222211]	L:3	E:15
[642222222222]	L:3	E:15
[633222222222]	L:3	E:15
[666321111111]	L:4	E:15
[665331111111]	L:4	E:15
[665222211111]	L:4	E:15
[664431111111]	L:4	E:15
[664422111111]	L:4	E:15
[664222221111]	L:4	E:15
[663332211111]	L:4	E:15
[663322221111]	L:4	E:15
[663222222111]	L:4	E:15
[655521111111]	L:4	E:15
[655222221111]	L:4	E:15
[654441111111]	L:4	E:15
[652222222221]	L:4	E:15
[644442111111]	L:4	E:15
[644433111111]	L:4	E:15
[644422221111]	L:4	E:15
[644222222211]	L:4	E:15
[643333311111]	L:4	E:15
[633333321111]	L:4	E:15
[633333222111]	L:4	E:15
[633332222211]	L:4	E:15
[633322222221]	L:4	E:15
[665421111111]	L:5	E:15
[665322111111]	L:5	E:15
[664332111111]	L:5	E:15
[664322211111]	L:5	E:15
[655431111111]	L:5	E:15
[655422111111]	L:5	E:15
[655332111111]	L:5	E:15
[655322211111]	L:5	E:15
[654422211111]	L:5	E:15
[654333111111]	L:5	E:15
[654222222111]	L:5	E:15
[653333211111]	L:5	E:15
[653332221111]	L:5	E:15
[653322222111]	L:5	E:15
[653222222211]	L:5	E:15
[644432211111]	L:5	E:15
[644333211111]	L:5	E:15
[644332221111]	L:5	E:15
[644322222111]	L:5	E:15
[643333221111]	L:5	E:15
[643332222111]	L:5	E:15
[643322222211]	L:5	E:15
[643222222221]	L:5	E:15
[654432111111]	L:6	E:15
[654332211111]	L:6	E:15
[654322221111]	L:6	E:15
[552222222222]	L:2	E:15
[555531111111]	L:3	E:15
[555522111111]	L:3	E:15
[555441111111]	L:3	E:15
[555333111111]	L:3	E:15
[555222222111]	L:3	E:15
[553333311111]	L:3	E:15
[533333331111]	L:3	E:15
[533322222222]	L:3	E:15
[555422211111]	L:4	E:15

[555332211111]	L:4	E:15
[555322221111]	L:4	E:15
[554442111111]	L:4	E:15
[554433111111]	L:4	E:15
[554422221111]	L:4	E:15
[554222222211]	L:4	E:15
[553333221111]	L:4	E:15
[553332222111]	L:4	E:15
[553322222211]	L:4	E:15
[553222222221]	L:4	E:15
[544443111111]	L:4	E:15
[544442211111]	L:4	E:15
[544422222111]	L:4	E:15
[544333311111]	L:4	E:15
[544222222221]	L:4	E:15
[543222222222]	L:4	E:15
[533333322111]	L:4	E:15
[533333222211]	L:4	E:15
[533332222221]	L:4	E:15
[555432111111]	L:5	E:15
[554432211111]	L:5	E:15
[554333211111]	L:5	E:15
[554332221111]	L:5	E:15
[554322222111]	L:5	E:15
[544433211111]	L:5	E:15
[544432221111]	L:5	E:15
[544333221111]	L:5	E:15
[544332222111]	L:5	E:15
[544322222211]	L:5	E:15
[543333321111]	L:5	E:15
[543333222111]	L:5	E:15
[543332222211]	L:5	E:15
[543322222221]	L:5	E:15
[444444111111]	L:2	E:15
[444222222222]	L:2	E:15
[444442221111]	L:3	E:15
[444433311111]	L:3	E:15
[444422222211]	L:3	E:15
[443333331111]	L:3	E:15
[443322222222]	L:3	E:15
[433332222222]	L:3	E:15
[444443211111]	L:4	E:15
[444433221111]	L:4	E:15
[444432222111]	L:4	E:15
[444333321111]	L:4	E:15
[444333222111]	L:4	E:15
[444332222211]	L:4	E:15
[444322222221]	L:4	E:15
[443333322111]	L:4	E:15
[443333222211]	L:4	E:15
[443332222221]	L:4	E:15
[433333332111]	L:4	E:15
[433333322211]	L:4	E:15
[433333222221]	L:4	E:15
[333333333111]	L:2	E:15
[333333222222]	L:2	E:15
[333333332211]	L:3	E:15
[333333322221]	L:3	E:15

Factor-16 Sequences

[B62222211111]	L:4	E:16
[B53322111111]	L:5	E:16
[B53222211111]	L:5	E:16
[B52222221111]	L:4	E:16
[B44331111111]	L:4	E:16
[B44322111111]	L:5	E:16
[B44222211111]	L:4	E:16
[B43332111111]	L:5	E:16
[B43322211111]	L:5	E:16
[B43222221111]	L:5	E:16
[B42222222111]	L:4	E:16
[B33333111111]	L:3	E:16
[B33332211111]	L:4	E:16
[B33322221111]	L:4	E:16
[B33222222111]	L:4	E:16
[B32222222211]	L:4	E:16
[B22222222221]	L:3	E:16
[A72222211111]	L:3	E:16
[A63322111111]	L:4	E:16
[A63222211111]	L:4	E:16
[A62222221111]	L:3	E:16
[A54331111111]	L:4	E:16
[A54322111111]	L:5	E:16
[A54222211111]	L:4	E:16
[A53332111111]	L:4	E:16

[A53322211111]	L:4	E:16
[A53222221111]	L:4	E:16
[A52222222111]	L:3	E:16
[A44431111111]	L:3	E:16
[A44422111111]	L:3	E:16
[A44332111111]	L:4	E:16
[A44322211111]	L:4	E:16
[A44222221111]	L:3	E:16
[A43333111111]	L:3	E:16
[A43332211111]	L:4	E:16
[A43322221111]	L:4	E:16
[A43222222111]	L:4	E:16
[A42222222211]	L:3	E:16
[A33333211111]	L:3	E:16
[A33332221111]	L:3	E:16
[A33322222111]	L:3	E:16
[A33222222211]	L:3	E:16
[A32222222221]	L:4	E:16
[A22222222222]	L:2	E:16
[982222211111]	L:4	E:16
[973322111111]	L:5	E:16
[973222211111]	L:5	E:16
[972222221111]	L:4	E:16
[964331111111]	L:5	E:16
[964322111111]	L:6	E:16
[964222211111]	L:5	E:16
[963332111111]	L:5	E:16
[963322211111]	L:5	E:16
[963222221111]	L:5	E:16
[962222222111]	L:4	E:16
[955331111111]	L:4	E:16
[955322111111]	L:5	E:16
[955222211111]	L:4	E:16
[954431111111]	L:5	E:16
[954422111111]	L:5	E:16
[954332111111]	L:6	E:16
[954322211111]	L:6	E:16
[954222221111]	L:5	E:16
[953333111111]	L:4	E:16
[953332211111]	L:5	E:16
[953322221111]	L:5	E:16
[953222222111]	L:5	E:16
[952222222211]	L:4	E:16
[944441111111]	L:3	E:16
[944432111111]	L:5	E:16
[944422211111]	L:4	E:16
[944333111111]	L:4	E:16
[944332211111]	L:5	E:16
[944322221111]	L:5	E:16
[944222222111]	L:4	E:16
[943333211111]	L:5	E:16
[943332221111]	L:5	E:16
[943322222111]	L:5	E:16
[943222222211]	L:5	E:16
[942222222221]	L:4	E:16
[933333311111]	L:3	E:16
[933333221111]	L:4	E:16
[933332222111]	L:4	E:16
[933322222211]	L:4	E:16
[933222222221]	L:4	E:16
[932222222222]	L:3	E:16
[883322111111]	L:3	E:16
[883222211111]	L:3	E:16
[882222221111]	L:2	E:16
[874331111111]	L:4	E:16
[874322111111]	L:5	E:16
[874222211111]	L:4	E:16
[873332111111]	L:4	E:16
[873322211111]	L:4	E:16
[873222221111]	L:4	E:16
[872222222111]	L:3	E:16
[865331111111]	L:4	E:16
[865322111111]	L:5	E:16
[865222211111]	L:4	E:16
[864431111111]	L:4	E:16
[864422111111]	L:4	E:16
[864332111111]	L:5	E:16
[864322211111]	L:5	E:16
[864222221111]	L:4	E:16
[863333111111]	L:3	E:16
[863332211111]	L:4	E:16
[863322221111]	L:4	E:16
[863222222111]	L:4	E:16
[862222222211]	L:3	E:16
[855431111111]	L:4	E:16
[855422111111]	L:4	E:16

[855332111111]	L:4	E:16
[855322211111]	L:4	E:16
[855222221111]	L:3	E:16
[854441111111]	L:3	E:16
[854432111111]	L:5	E:16
[854422211111]	L:4	E:16
[854333111111]	L:4	E:16
[854332211111]	L:5	E:16
[854322221111]	L:5	E:16
[854222222111]	L:4	E:16
[853333211111]	L:4	E:16
[853332221111]	L:4	E:16
[853322222111]	L:4	E:16
[853222222211]	L:4	E:16
[852222222221]	L:4	E:16
[844442111111]	L:3	E:16
[844433111111]	L:3	E:16
[844432211111]	L:4	E:16
[844422221111]	L:3	E:16
[844333211111]	L:4	E:16
[844332221111]	L:4	E:16
[844322222111]	L:4	E:16
[844222222211]	L:3	E:16
[843333311111]	L:3	E:16
[843333221111]	L:4	E:16
[843332222111]	L:4	E:16
[843322222211]	L:4	E:16
[843222222221]	L:5	E:16
[842222222222]	L:3	E:16
[833333321111]	L:3	E:16
[833333222111]	L:3	E:16
[833332222211]	L:3	E:16
[833322222221]	L:4	E:16
[833222222222]	L:3	E:16
[775331111111]	L:4	E:16
[775322111111]	L:5	E:16
[775222211111]	L:4	E:16
[774431111111]	L:4	E:16
[774422111111]	L:4	E:16
[774332111111]	L:5	E:16
[774322211111]	L:5	E:16
[774222221111]	L:4	E:16
[773333111111]	L:3	E:16
[773332211111]	L:4	E:16
[773322221111]	L:4	E:16
[773222222111]	L:4	E:16
[772222222211]	L:3	E:16
[766331111111]	L:4	E:16
[766322111111]	L:5	E:16
[766222211111]	L:4	E:16
[765431111111]	L:6	E:16
[765422111111]	L:6	E:16
[765332111111]	L:6	E:16
[765322211111]	L:6	E:16
[765222221111]	L:5	E:16
[764441111111]	L:4	E:16
[764432111111]	L:6	E:16
[764422211111]	L:5	E:16
[764333111111]	L:5	E:16
[764332211111]	L:6	E:16
[764322221111]	L:6	E:16
[764222222111]	L:5	E:16
[763333211111]	L:5	E:16
[763332221111]	L:5	E:16
[763322222111]	L:5	E:16
[763222222211]	L:5	E:16
[762222222221]	L:4	E:16
[755531111111]	L:4	E:16
[755522111111]	L:4	E:16
[755441111111]	L:4	E:16
[755432111111]	L:6	E:16
[755422211111]	L:5	E:16
[755333111111]	L:4	E:16
[755332211111]	L:5	E:16
[755322221111]	L:5	E:16
[755222222111]	L:4	E:16
[754442111111]	L:5	E:16
[754433111111]	L:5	E:16
[754432211111]	L:6	E:16
[754422221111]	L:5	E:16
[754333211111]	L:6	E:16
[754332221111]	L:6	E:16
[754322222111]	L:6	E:16
[754222222211]	L:5	E:16
[753333311111]	L:4	E:16
[753333221111]	L:5	E:16

[753332222111]	L:5	E:16
[753322222211]	L:5	E:16
[753222222221]	L:5	E:16
[752222222222]	L:3	E:16
[744443111111]	L:4	E:16
[744442211111]	L:4	E:16
[744433211111]	L:5	E:16
[744432221111]	L:5	E:16
[744422222111]	L:4	E:16
[744333311111]	L:4	E:16
[744333221111]	L:5	E:16
[744332222111]	L:5	E:16
[744322222211]	L:5	E:16
[744222222221]	L:4	E:16
[743333321111]	L:5	E:16
[743333222111]	L:5	E:16
[743332222211]	L:5	E:16
[743322222221]	L:5	E:16
[743222222222]	L:4	E:16
[733333331111]	L:3	E:16
[733333322111]	L:4	E:16
[733333222211]	L:4	E:16
[733332222221]	L:4	E:16
[733322222222]	L:3	E:16
[666431111111]	L:4	E:16
[666422111111]	L:4	E:16
[666332111111]	L:4	E:16
[666322211111]	L:4	E:16
[666222221111]	L:3	E:16
[665531111111]	L:4	E:16
[665522111111]	L:4	E:16
[665441111111]	L:4	E:16
[665432111111]	L:6	E:16
[665422211111]	L:5	E:16
[665333111111]	L:4	E:16
[665332211111]	L:5	E:16
[665322221111]	L:5	E:16
[665222222111]	L:4	E:16
[664442111111]	L:4	E:16
[664433111111]	L:4	E:16
[664432211111]	L:5	E:16
[664422221111]	L:4	E:16
[664333211111]	L:5	E:16
[664332221111]	L:5	E:16
[664322222111]	L:5	E:16
[664222222211]	L:4	E:16
[663333311111]	L:3	E:16
[663333221111]	L:4	E:16
[663332222111]	L:4	E:16
[663322222211]	L:4	E:16
[663222222221]	L:4	E:16
[662222222222]	L:2	E:16
[655541111111]	L:4	E:16
[655532111111]	L:5	E:16
[655522211111]	L:4	E:16
[655442111111]	L:5	E:16
[655433111111]	L:5	E:16
[655432211111]	L:6	E:16
[655422221111]	L:5	E:16
[655333211111]	L:5	E:16
[655332221111]	L:5	E:16
[655322222111]	L:5	E:16
[655222222211]	L:4	E:16
[654443111111]	L:5	E:16
[654442211111]	L:5	E:16
[654433211111]	L:6	E:16
[654432221111]	L:6	E:16
[654422222111]	L:5	E:16
[654333311111]	L:5	E:16
[654333221111]	L:6	E:16
[654332222111]	L:6	E:16
[654322222211]	L:6	E:16
[654222222221]	L:5	E:16
[653333321111]	L:5	E:16
[653333222111]	L:5	E:16
[653332222211]	L:5	E:16
[653322222221]	L:5	E:16
[653222222222]	L:4	E:16
[644444111111]	L:3	E:16
[644443211111]	L:5	E:16
[644442221111]	L:4	E:16
[644433311111]	L:4	E:16
[644433221111]	L:5	E:16
[644432222111]	L:5	E:16
[644422222211]	L:4	E:16
[644333321111]	L:5	E:16

[644333222111] L:5 E:16
[644332222211] L:5 E:16
[644322222221] L:5 E:16
[644222222222] L:3 E:16
[643333331111] L:4 E:16
[643333322111] L:5 E:16
[643333222211] L:5 E:16
[643332222221] L:5 E:16
[643322222222] L:4 E:16
[633333332111] L:4 E:16
[633333322211] L:4 E:16
[633333222221] L:4 E:16
[633332222222] L:3 E:16
[555551111111] L:2 E:16
[555542111111] L:4 E:16
[555533111111] L:3 E:16
[555532211111] L:4 E:16
[555522221111] L:3 E:16
[555443111111] L:4 E:16
[555442211111] L:4 E:16
[555433211111] L:5 E:16
[555432221111] L:5 E:16
[555422222111] L:4 E:16
[555333311111] L:3 E:16
[555333221111] L:4 E:16
[555332222111] L:4 E:16
[555322222211] L:4 E:16
[555222222221] L:3 E:16
[554444111111] L:3 E:16
[554443211111] L:5 E:16
[554442221111] L:4 E:16
[554433311111] L:4 E:16
[554433221111] L:5 E:16
[554432222111] L:5 E:16
[554422222211] L:4 E:16
[554333321111] L:5 E:16
[554333222111] L:5 E:16
[554332222211] L:5 E:16
[554322222221] L:5 E:16
[554222222222] L:3 E:16
[553333331111] L:3 E:16
[553333322111] L:4 E:16
[553333222211] L:4 E:16
[553332222221] L:4 E:16
[553322222222] L:3 E:16
[544444211111] L:4 E:16
[544443311111] L:4 E:16
[544443221111] L:5 E:16
[544442222111] L:4 E:16
[544433321111] L:5 E:16
[544433222111] L:5 E:16
[544432222211] L:5 E:16
[544422222221] L:4 E:16
[544333331111] L:4 E:16
[544333322111] L:5 E:16
[544333222211] L:5 E:16
[544332222221] L:5 E:16
[544322222222] L:4 E:16
[543333332111] L:5 E:16
[543333322211] L:5 E:16
[543333222221] L:5 E:16
[543332222222] L:4 E:16
[533333333111] L:3 E:16
[533333332211] L:4 E:16
[533333322221] L:4 E:16
[533333222222] L:3 E:16
[444444311111] L:3 E:16
[444444221111] L:3 E:16
[444443321111] L:4 E:16
[444443222111] L:4 E:16
[444442222211] L:3 E:16
[444433331111] L:3 E:16
[444433322111] L:4 E:16
[444433222211] L:4 E:16
[444432222221] L:4 E:16
[444422222222] L:2 E:16
[444333332111] L:4 E:16
[444333322211] L:4 E:16
[444333222221] L:4 E:16
[444332222222] L:3 E:16
[443333333111] L:3 E:16
[443333332211] L:4 E:16
[443333322221] L:4 E:16
[443333222222] L:3 E:16
[433333333211] L:4 E:16
[433333332221] L:4 E:16

[433333322222]	L:3	E:16
[333333333311]	L:2	E:16
[333333333221]	L:3	E:16
[333333332222]	L:2	E:16

Factor-17 Sequences

[B72222221111]	L:4	E:17
[B63322211111]	L:5	E:17
[B63222221111]	L:5	E:17
[B62222222111]	L:4	E:17
[B54332111111]	L:6	E:17
[B54322211111]	L:6	E:17
[B54222221111]	L:5	E:17
[B53333111111]	L:4	E:17
[B53332211111]	L:5	E:17
[B53322221111]	L:5	E:17
[B53222222111]	L:5	E:17
[B52222222211]	L:4	E:17
[B44441111111]	L:3	E:17
[B44432111111]	L:5	E:17
[B44422211111]	L:4	E:17
[B44333111111]	L:4	E:17
[B44332211111]	L:5	E:17
[B44322221111]	L:5	E:17
[B44222222111]	L:4	E:17
[B43333211111]	L:5	E:17
[B43332221111]	L:5	E:17
[B43322222111]	L:5	E:17
[B43222222211]	L:5	E:17
[B42222222221]	L:4	E:17
[B33333311111]	L:3	E:17
[B33333221111]	L:4	E:17
[B33332222111]	L:4	E:17
[B33322222211]	L:4	E:17
[B33222222221]	L:4	E:17
[B32222222222]	L:3	E:17
[A82222221111]	L:3	E:17
[A82222221111]	L:3	E:17
[A73322211111]	L:4	E:17
[A73222221111]	L:4	E:17
[A72222222111]	L:3	E:17
[A64332111111]	L:5	E:17
[A64322211111]	L:5	E:17
[A64222221111]	L:4	E:17
[A63333111111]	L:3	E:17
[A63332211111]	L:4	E:17
[A63322221111]	L:4	E:17
[A63222222111]	L:4	E:17
[A62222222211]	L:3	E:17
[A55332111111]	L:4	E:17
[A55322211111]	L:4	E:17
[A55222221111]	L:3	E:17
[A54441111111]	L:3	E:17
[A54432111111]	L:5	E:17
[A54422211111]	L:4	E:17
[A54333111111]	L:4	E:17
[A54332211111]	L:5	E:17
[A54322221111]	L:5	E:17
[A54222222111]	L:4	E:17
[A53333211111]	L:4	E:17
[A53332221111]	L:4	E:17
[A53322222111]	L:4	E:17
[A53222222211]	L:4	E:17
[A52222222221]	L:4	E:17
[A44442111111]	L:3	E:17
[A44433111111]	L:3	E:17
[A44432211111]	L:4	E:17
[A44422221111]	L:3	E:17
[A44333211111]	L:4	E:17
[A44332221111]	L:4	E:17
[A44322222111]	L:4	E:17
[A44222222211]	L:3	E:17
[A43333311111]	L:3	E:17
[A43333221111]	L:4	E:17
[A43332222111]	L:4	E:17
[A43322222211]	L:4	E:17
[A43222222221]	L:5	E:17
[A42222222222]	L:3	E:17
[A33333321111]	L:3	E:17
[A33333222111]	L:3	E:17
[A33332222211]	L:3	E:17
[A33322222221]	L:4	E:17
[A33222222222]	L:3	E:17
[992222221111]	L:3	E:17
[983322211111]	L:5	E:17

[983222221111]	L:5	E:17
[982222222111]	L:4	E:17
[974332111111]	L:6	E:17
[974322211111]	L:6	E:17
[974222221111]	L:5	E:17
[973333111111]	L:4	E:17
[973332211111]	L:5	E:17
[973322221111]	L:5	E:17
[973222222111]	L:5	E:17
[972222222211]	L:4	E:17
[965332111111]	L:6	E:17
[965322211111]	L:6	E:17
[965222221111]	L:5	E:17
[964441111111]	L:4	E:17
[964432111111]	L:6	E:17
[964422211111]	L:5	E:17
[964333111111]	L:5	E:17
[964332211111]	L:6	E:17
[964322221111]	L:6	E:17
[964222222111]	L:5	E:17
[963333211111]	L:5	E:17
[963332221111]	L:5	E:17
[963322222111]	L:5	E:17
[963222222211]	L:5	E:17
[962222222221]	L:4	E:17
[955441111111]	L:4	E:17
[955432111111]	L:6	E:17
[955422211111]	L:5	E:17
[955333111111]	L:4	E:17
[955332211111]	L:5	E:17
[955322221111]	L:5	E:17
[955222222111]	L:4	E:17
[954442111111]	L:5	E:17
[954433111111]	L:5	E:17
[954432211111]	L:6	E:17
[954422221111]	L:5	E:17
[954333211111]	L:6	E:17
[954332221111]	L:6	E:17
[954322222111]	L:6	E:17
[954222222211]	L:5	E:17
[953333311111]	L:4	E:17
[953333221111]	L:5	E:17
[953332222111]	L:5	E:17
[953322222211]	L:5	E:17
[953222222221]	L:5	E:17
[952222222222]	L:3	E:17
[944443111111]	L:4	E:17
[944442211111]	L:4	E:17
[944433211111]	L:5	E:17
[944432221111]	L:5	E:17
[944422222111]	L:4	E:17
[944333311111]	L:4	E:17
[944333221111]	L:5	E:17
[944332222111]	L:5	E:17
[944322222211]	L:5	E:17
[944222222221]	L:4	E:17
[943333321111]	L:5	E:17
[943333222111]	L:5	E:17
[943332222211]	L:5	E:17
[943322222221]	L:5	E:17
[943222222222]	L:4	E:17
[933333331111]	L:3	E:17
[933333322111]	L:4	E:17
[933333222211]	L:4	E:17
[933332222221]	L:4	E:17
[933322222222]	L:3	E:17
[884332111111]	L:4	E:17
[884322211111]	L:4	E:17
[884222221111]	L:3	E:17
[883333111111]	L:2	E:17
[883332211111]	L:3	E:17
[883322221111]	L:3	E:17
[883222222111]	L:3	E:17
[882222222211]	L:2	E:17
[875332111111]	L:5	E:17
[875322211111]	L:5	E:17
[875222221111]	L:4	E:17
[874441111111]	L:3	E:17
[874432111111]	L:5	E:17
[874422211111]	L:4	E:17
[874333111111]	L:4	E:17
[874332211111]	L:5	E:17
[874322221111]	L:5	E:17
[874222222111]	L:4	E:17
[873333211111]	L:4	E:17
[873332221111]	L:4	E:17

[873322222111]	L:4	E:17
[873222222211]	L:4	E:17
[872222222221]	L:4	E:17
[866332111111]	L:4	E:17
[866322211111]	L:4	E:17
[866222221111]	L:3	E:17
[865441111111]	L:4	E:17
[865432111111]	L:6	E:17
[865422211111]	L:5	E:17
[865333111111]	L:4	E:17
[865332211111]	L:5	E:17
[865322221111]	L:5	E:17
[865222222111]	L:4	E:17
[864442111111]	L:4	E:17
[864433111111]	L:4	E:17
[864432211111]	L:5	E:17
[864422221111]	L:4	E:17
[864333211111]	L:5	E:17
[864332221111]	L:5	E:17
[864322222111]	L:5	E:17
[864222222211]	L:4	E:17
[863333311111]	L:3	E:17
[863333221111]	L:4	E:17
[863332222111]	L:4	E:17
[863322222211]	L:4	E:17
[863222222221]	L:5	E:17
[862222222222]	L:3	E:17
[855541111111]	L:3	E:17
[855532111111]	L:4	E:17
[855522211111]	L:3	E:17
[855442111111]	L:4	E:17
[855433111111]	L:4	E:17
[855432211111]	L:5	E:17
[855422221111]	L:4	E:17
[855333211111]	L:4	E:17
[855332221111]	L:4	E:17
[855322222111]	L:4	E:17
[855222222211]	L:3	E:17
[854443111111]	L:4	E:17
[854442211111]	L:4	E:17
[854433211111]	L:5	E:17
[854432221111]	L:5	E:17
[854422222111]	L:4	E:17
[854333311111]	L:4	E:17
[854333221111]	L:5	E:17
[854332222111]	L:5	E:17
[854322222211]	L:5	E:17
[854222222221]	L:5	E:17
[853333321111]	L:4	E:17
[853333222111]	L:4	E:17
[853332222211]	L:4	E:17
[853322222221]	L:5	E:17
[853222222222]	L:4	E:17
[844444111111]	L:2	E:17
[844443211111]	L:4	E:17
[844442221111]	L:3	E:17
[844433311111]	L:3	E:17
[844433221111]	L:4	E:17
[844432222111]	L:4	E:17
[844422222211]	L:3	E:17
[844333321111]	L:4	E:17
[844333222111]	L:4	E:17
[844332222211]	L:4	E:17
[844322222221]	L:5	E:17
[844222222222]	L:3	E:17
[843333331111]	L:3	E:17
[843333322111]	L:4	E:17
[843333222211]	L:4	E:17
[843332222221]	L:5	E:17
[843322222222]	L:4	E:17
[833333332111]	L:3	E:17
[833333322211]	L:3	E:17
[833333222221]	L:4	E:17
[833332222222]	L:3	E:17
[776332111111]	L:5	E:17
[776322211111]	L:5	E:17
[776222221111]	L:4	E:17
[775441111111]	L:4	E:17
[775432111111]	L:6	E:17
[775422211111]	L:5	E:17
[775333111111]	L:4	E:17
[775332211111]	L:5	E:17
[775322221111]	L:5	E:17
[775222222111]	L:4	E:17
[774442111111]	L:4	E:17
[774433111111]	L:4	E:17

[774432211111]	L:5	E:17
[774422221111]	L:4	E:17
[774333211111]	L:5	E:17
[774332221111]	L:5	E:17
[774322222111]	L:5	E:17
[774222222211]	L:4	E:17
[773333311111]	L:3	E:17
[773333221111]	L:4	E:17
[773332222111]	L:4	E:17
[773322222211]	L:4	E:17
[773222222221]	L:4	E:17
[772222222222]	L:2	E:17
[766441111111]	L:4	E:17
[766432111111]	L:6	E:17
[766422211111]	L:5	E:17
[766333111111]	L:4	E:17
[766332211111]	L:5	E:17
[766322221111]	L:5	E:17
[766222222111]	L:4	E:17
[765541111111]	L:5	E:17
[765532111111]	L:6	E:17
[765522211111]	L:5	E:17
[765442111111]	L:6	E:17
[765433111111]	L:6	E:17
[765432211111]	L:7	E:17
[765422221111]	L:6	E:17
[765333211111]	L:6	E:17
[765332221111]	L:6	E:17
[765322222111]	L:6	E:17
[765222222211]	L:5	E:17
[764444311111]	L:5	E:17
[764444221111]	L:5	E:17
[764433211111]	L:6	E:17
[764432221111]	L:6	E:17
[764422222111]	L:5	E:17
[764333311111]	L:5	E:17
[764333221111]	L:6	E:17
[764332222111]	L:6	E:17
[764322222211]	L:6	E:17
[764222222221]	L:5	E:17
[763333321111]	L:5	E:17
[763333222111]	L:5	E:17
[763332222211]	L:5	E:17
[763322222221]	L:5	E:17
[763222222222]	L:4	E:17
[755551111111]	L:3	E:17
[755542111111]	L:5	E:17
[755533111111]	L:4	E:17
[755532211111]	L:5	E:17
[755522221111]	L:4	E:17
[755443111111]	L:5	E:17
[755442211111]	L:5	E:17
[755433211111]	L:6	E:17
[755432221111]	L:6	E:17
[755422222111]	L:5	E:17
[755333311111]	L:4	E:17
[755333221111]	L:5	E:17
[755332222111]	L:5	E:17
[755322222211]	L:5	E:17
[755222222221]	L:4	E:17
[754444111111]	L:4	E:17
[754443211111]	L:6	E:17
[754442221111]	L:5	E:17
[754433311111]	L:5	E:17
[754433221111]	L:6	E:17
[754432222111]	L:6	E:17
[754422222211]	L:5	E:17
[754333321111]	L:6	E:17
[754333222111]	L:6	E:17
[754332222211]	L:6	E:17
[754322222221]	L:6	E:17
[754222222222]	L:4	E:17
[753333331111]	L:4	E:17
[753333322111]	L:5	E:17
[753333222211]	L:5	E:17
[753332222221]	L:5	E:17
[753322222222]	L:4	E:17
[744444211111]	L:4	E:17
[744443311111]	L:4	E:17
[744443221111]	L:5	E:17
[744442222111]	L:4	E:17
[744433321111]	L:5	E:17
[744433222111]	L:5	E:17
[744432222211]	L:5	E:17
[744422222221]	L:4	E:17
[744333331111]	L:4	E:17

[744333322111]	L:5	E:17
[744333222211]	L:5	E:17
[744332222221]	L:5	E:17
[744322222222]	L:4	E:17
[743333332111]	L:5	E:17
[743333322211]	L:5	E:17
[743333222221]	L:5	E:17
[743332222222]	L:4	E:17
[733333333111]	L:3	E:17
[733333332211]	L:4	E:17
[733333322221]	L:4	E:17
[733333222222]	L:3	E:17
[666541111111]	L:4	E:17
[666532111111]	L:5	E:17
[666522211111]	L:4	E:17
[666442111111]	L:4	E:17
[666433111111]	L:4	E:17
[666432211111]	L:5	E:17
[666422221111]	L:4	E:17
[666333211111]	L:4	E:17
[666332221111]	L:4	E:17
[666322222111]	L:4	E:17
[666222222211]	L:3	E:17
[665551111111]	L:3	E:17
[665542111111]	L:5	E:17
[665533111111]	L:4	E:17
[665532211111]	L:5	E:17
[665522221111]	L:4	E:17
[665443111111]	L:5	E:17
[665442211111]	L:5	E:17
[665433211111]	L:6	E:17
[665432221111]	L:6	E:17
[665422222111]	L:5	E:17
[665333311111]	L:4	E:17
[665333221111]	L:5	E:17
[665332222111]	L:5	E:17
[665322222211]	L:5	E:17
[665222222221]	L:4	E:17
[664444111111]	L:3	E:17
[664443211111]	L:5	E:17
[664442221111]	L:4	E:17
[664433311111]	L:4	E:17
[664433221111]	L:5	E:17
[664432222111]	L:5	E:17
[664422222211]	L:4	E:17
[664333321111]	L:5	E:17
[664333222111]	L:5	E:17
[664332222211]	L:5	E:17
[664322222221]	L:5	E:17
[664222222222]	L:3	E:17
[663333331111]	L:3	E:17
[663333322111]	L:4	E:17
[663333222211]	L:4	E:17
[663332222221]	L:4	E:17
[663322222222]	L:3	E:17
[655552111111]	L:4	E:17
[655543111111]	L:5	E:17
[655542211111]	L:5	E:17
[655533211111]	L:5	E:17
[655532221111]	L:5	E:17
[655522222111]	L:4	E:17
[655444111111]	L:4	E:17
[655443211111]	L:6	E:17
[655442221111]	L:5	E:17
[655433311111]	L:5	E:17
[655433221111]	L:6	E:17
[655432222111]	L:6	E:17
[655422222211]	L:5	E:17
[655333321111]	L:5	E:17
[655333222111]	L:5	E:17
[655332222211]	L:5	E:17
[655322222221]	L:5	E:17
[655222222222]	L:3	E:17
[654444211111]	L:5	E:17
[654443311111]	L:5	E:17
[654443221111]	L:6	E:17
[654442222111]	L:5	E:17
[654433321111]	L:6	E:17
[654433222111]	L:6	E:17
[654432222211]	L:6	E:17
[654422222221]	L:5	E:17
[654333331111]	L:5	E:17
[654333322111]	L:6	E:17
[654333222211]	L:6	E:17
[654332222221]	L:6	E:17
[654322222222]	L:5	E:17

[653333332111]	L:5	E:17
[653333322211]	L:5	E:17
[653333222221]	L:5	E:17
[653332222222]	L:4	E:17
[644444311111]	L:4	E:17
[644444221111]	L:4	E:17
[644443321111]	L:5	E:17
[644443222111]	L:5	E:17
[644442222211]	L:4	E:17
[644433331111]	L:4	E:17
[644433322111]	L:5	E:17
[644433222211]	L:5	E:17
[644432222221]	L:5	E:17
[644422222222]	L:3	E:17
[644333332111]	L:5	E:17
[644333322211]	L:5	E:17
[644333222221]	L:5	E:17
[644332222222]	L:4	E:17
[643333333111]	L:4	E:17
[643333332211]	L:5	E:17
[643333322221]	L:5	E:17
[643333222222]	L:4	E:17
[633333333211]	L:4	E:17
[633333332221]	L:4	E:17
[633333322222]	L:3	E:17
[555553111111]	L:3	E:17
[555552211111]	L:3	E:17
[555544111111]	L:3	E:17
[555543211111]	L:5	E:17
[555542221111]	L:4	E:17
[555533311111]	L:3	E:17
[555533221111]	L:4	E:17
[555532222111]	L:4	E:17
[555522222211]	L:3	E:17
[555444211111]	L:4	E:17
[555443311111]	L:4	E:17
[555443221111]	L:5	E:17
[555442222111]	L:4	E:17
[555433321111]	L:5	E:17
[555433222111]	L:5	E:17
[555432222211]	L:5	E:17
[555422222221]	L:4	E:17
[555333331111]	L:3	E:17
[555333322111]	L:4	E:17
[555333222211]	L:4	E:17
[555332222221]	L:4	E:17
[555322222222]	L:3	E:17
[554444311111]	L:4	E:17
[554444221111]	L:4	E:17
[554443321111]	L:5	E:17
[554443222111]	L:5	E:17
[554442222211]	L:4	E:17
[554433331111]	L:4	E:17
[554433322111]	L:5	E:17
[554433222211]	L:5	E:17
[554432222221]	L:5	E:17
[554422222222]	L:3	E:17
[554333332111]	L:5	E:17
[554333322211]	L:5	E:17
[554333222221]	L:5	E:17
[554332222222]	L:4	E:17
[553333333111]	L:3	E:17
[553333332211]	L:4	E:17
[553333322221]	L:4	E:17
[553333222222]	L:3	E:17
[544444411111]	L:3	E:17
[544444321111]	L:5	E:17
[544444222111]	L:4	E:17
[544443331111]	L:4	E:17
[544443322111]	L:5	E:17
[544443222211]	L:5	E:17
[544442222221]	L:4	E:17
[544433332111]	L:5	E:17
[544433322211]	L:5	E:17
[544433222221]	L:5	E:17
[544432222222]	L:4	E:17
[544333333111]	L:4	E:17
[544333332211]	L:5	E:17
[544333322221]	L:5	E:17
[544333222222]	L:4	E:17
[543333333211]	L:5	E:17
[543333332221]	L:5	E:17
[543333322222]	L:4	E:17
[533333333311]	L:3	E:17
[533333333221]	L:4	E:17
[533333332222]	L:3	E:17

[444444421111]	L:3	E:17
[444444331111]	L:3	E:17
[444444322111]	L:4	E:17
[444444222211]	L:3	E:17
[444443332111]	L:4	E:17
[444443322211]	L:4	E:17
[444443222221]	L:4	E:17
[444442222222]	L:2	E:17
[444433333111]	L:3	E:17
[444433332211]	L:4	E:17
[444433322221]	L:4	E:17

[444433222222]	L:3	E:17
[444333333211]	L:4	E:17
[444333332221]	L:4	E:17
[444333322222]	L:3	E:17
[443333333311]	L:3	E:17
[443333333221]	L:4	E:17
[443333332222]	L:3	E:17
[433333333321]	L:4	E:17
[433333333222]	L:3	E:17
[333333333331]	L:2	E:17
[333333333322]	L:2	E:17

Bibliography

Agmon, Eytan
1989 A mathematical model of the diatonic system. *Journal of Music Theory* 33/1:1-26.
1990 Equal divisions of the octave in a Scarlatti sonata. *In Theory Only* 11/5:1-8.

Alphonce, Bo
1974 *The Invariance Matrix.* Ph.D. dissertation, Yale University.
1982 Review of *Twelve-tone tonality*, by George Perle. *Journal of Music Theory* 26/1:179-205.
1988 Review of *Generalized musical intervals and transformations*, by David Lewin. *Intégral* 2:161-78.

Andrews, William Symes
1917 *Magic squares and cubes.* Chicago: Open Court Publishing Company.

Antokoletz, Elliott
1984 *The music of Béla Bartók.* Berkeley: University of California Press.
1986 Interval cycles in Stravinsky's early ballets. *Journal of the American Musicological Society* 39/3:579-89.
1992 *Twentieth-century music.* Englewood Cliffs, NJ: Prentice-Hall.

Arikati, Srinivasa
1994 *On the parallel complexity of degree sequence problems.* Saarbrucken, Germany: Max-Planck-Institut für Informatik.

Atlas, Raphael and Cherlin, Michael, eds.
1994 *Musical transformation and musical intuition: Eleven essays in honor of David Lewin.* Roxbury, MA: Ovenbird Press.

Babbitt, Milton
1952 Review of *Structural hearing,* by Felix Salzer. *Journal of the American Musicological Society* 5/3:260-5.
1953 Review of *The life and music of Béla Bartók*, by Halsley Stevens. *Saturday Review* 36/30:23.
1954 Musical America's several generations. *Saturday Review* 37/11:36.
1955 Some aspects of twelve-tone composition. *The Score and IMA Magazine* 12:53-61.

1958 Who cares if you listen? *High Fidelity Magazine* 8/2:38-40. Also in Schwartz and Childs (1967:243-50).
1960 Twelve-tone invariants as compositional determinants. *Musical Quarterly* 46:245-59.
1961a Set structure as a compositional determinant. *Journal of Music Theory* 5/2:72-94. Also in Boretz and Cone (1972a:129-47).
1961b Past and present concepts of the nature and limits of music. *International Musicological Society Congress Report* Boretz and Cone (1972a:3-9).
1961c Set structure as a compositional determinant. *Journal of Music Theory* 5/2:72-94.
1962 Twelve-tone rhythmic structure and the electronic medium. *Perspectives of New Music* 1/1:49-79. Also in Boretz and Cone (1972a:148-79).
1963 Mr. Babbitt answers. *Perspectives of New Music* 2/1:127-32 (See Perle 1963).
1964a Remarks on the recent Stravinsky. *Perspectives of New Music* 2/2:35-55.
1964b The synthesis, perception, and specification of musical time. *Journal of the International Folk Music Council* 16:92-95.
1965 The structure and function of musical theory. *College Music Symposium* 5:49-60. Also in Boretz and Cone (1972a:10-21).
1966 Edgard Varèse: A few observations on his music. *Perspectives of New Music* 4/2:14-22.
1968 Three essays on Schoenberg: *Concerto for Violin and Orchestra, Das Buch der hängenden Gärten, Moses and Aron.* Boretz and Cone (1972b:47-60).
1970 On *Relata 1. Perspectives of New Music* 9/1:1-22.
1971 Contemporary music composition and music theory as contemporary intellectual history. In *Perspectives in Musicology*, ed. B. Brooks, E. Downes, S. Van Solkema. New York: W.W. Norton:151-84.
1973 Since Schoenberg. *Perspectives of New Music* 12/1:3-28.
1976 Responses: A first approximation. *Perspectives of New Music* 15/1:3-23.
1976b Letter (On Mallalieu row). *In Theory Only* 2/7:9.
1986 Order, symmetry, and centricity. In *Confronting Stravinsky* (Pasler 1986:247-61).

Bailey, Kathryn
1991 *The Twelve-note music of Anton Webern; Old forms in a new language.* Cambridge: Cambridge University Press.

Bailey, Marshall
1962 Duodecuple notation. *American Composers Alliance Bulletin* 10/3:12-14.

Basert, Ann Phillips
1961 *Serial Music: A Classified bibliography on 12-tone and electronic music.* Berkeley: University of California Press.

Bass, Richard
1991 Sets, scales, and symmetries: The pitch-structural basis of George Crumb's *Makrokosmos I and II. Music Theory Spectrum* 13/1:1-20.

Batstone, Philip
1972 Multiple-order functions in twelve-tone music. *Perspectives of New Music* 10/2: 60-71, 11/1:92-111.

Bauer, Douglas
1980 Line-graphical degree sequences. *Journal of Graph Theory* 4:219-232.

Bauer-Mengelberg, Stefan and Ferentz, Melvin
1965 On eleven-interval twelve-tone rows. *Perspectives of New Music* 3/2:93-103.

Bazelow, Alexander R. and Brickle, Frank
1976 A partition problem posed by Milton Babbitt (Part 1). *Perspectives of New Music* 15/1:280-93.
1977 A combinatorial problem arising out of twelve-tone theory. *In Theory Only* 3/4:12-13.

Beach, David W.
1976 Segmental invariance and the twelve-tone system. *Journal of Music Theory* 20/2:157-84.
1979 Pitch structure and the analytic process in atonal music: An interpretation of the theory of sets. *Music Theory Spectrum* 1:7-22.
Behzad, Mehdi and Chartrand, Gary
1971 *Introduction to the theory of graphs.* Boston: Allyn and Bacon.
Benjamin, William
1974 Review of *The structure of atonal music*, by Allen Forte. *Perspectives of New Music* 13/1:170-90.
1975 Towards a typology for pitch equivalence. *In Theory Only* 1/6:3-8.
1976 On pitch relationships: Unmasking some myths about octave equivalence. *Proceedings of the American Society of University Composers* 11/1:28-37.
Bennighof, James
1987 Set-class aggregate structuring, graph theory, and some compositional strategies. *Journal of Music Theory* 31/1:51-98.
Berger, Arthur
1963 Problems of pitch organization in Stravinsky. *Perspectives of New Music* 2/1:11-42. Also in Boretz and Cone (1971).
1964 New linguistic modes and the new theory. *Perspectives of New Music* 3/1:1-9. Also in Boretz and Cone (1972).
Berman, Gerald, ed.
1978 *Applied graph theory bibliography: with foreword citations.* Waterloo, Ont.: University of Waterloo.
Bernard, Jonathan W.
1981 Pitch/register in the music of Edgard Varèse. *Music Theory Spectrum* 3:1-25.
1983 Spatial sets in recent music of Elliott Carter. *Music Analysis* 2/1:5-34.
1986a Messiaen's synaesthesia: The correspondence between color and sound structure in his music. *Music Perception* 4/1:41-68.
1987 *The Music of Edgard Varèse.* New Haven: Yale University Press.
1988 The evolution of Elliott Carter's rhythmic practice. *Perspectives of New Music* 26/2:164-203.
1992 Cracked octaves, warped perspectives: A response to Guck (1992). *Perspectives of New Music* 30/2:274-89.
1994 Review of *The listening composer*, by George Perle. *Journal of Music Theory* 38/1:103-122.
Berry, Wallace
1976 *Structural functions in music.* New York: Dover Publications Reprinted:1987.
1980 Symmetrical interval sets and derivative pitch materials in Bartók's *String Quartet* No.3. *Perspectives of New Music* 18/1-2:287-380.
1980 On structural levels in music. *Music Theory Spectrum* 2:19-45.
Besant, Annie, and Leadbeater, Charles Webster
1905 *Thought-forms.* London: Theosophical Publishing House.
Biggs, Norman
1974 *Algebraic graph theory.* London: Cambridge University Press.
1985 *Discrete mathematics.* Oxford: Clarendon Press.
Biggs, Norman, Lloyd, E. Keith, and Wilson, Richard J.
1976 *Graph theory 1736-1936.* Oxford: Clarendon Press.
Block, Steven
1990 Pitch-class transformation in free jazz. *Music Theory Spectrum* 12/2:181-202.
Bollabás, Béla
1978 *Extremal graph theory.* London: Academic Press.

Bonchev, Danail and Rouvray, Dennis, eds.
1991 *Chemical graph theory; Introduction and fundamentals.* New York: Abacus Press.
Bondy, John Adrian and Murty, U. S. R.
1976 *Graph theory with applications.* New York: North-Holland.
Boretz, Benjamin
1963 Review of *Serial composition and atonality*, by George Perle. *Perspectives of New Music* 1/2:125-36.
1969 Meta-variations: Studies in the foundations of musical thought. *Perspectives of New Music* 8/1:1-75.
1970a Sketch of a musical system (Meta-Variations, Part II). *Perspectives of New Music* 8/2:49-112.
1970b The construction of musical syntax. *Perspectives of New Music* 9/1:23-42.
1971a Musical syntax (II). *Perspectives of New Music* 9/2-10/1:232-70.
1971b In quest of the rhythmic genius. *Perspectives of New Music* 9/2-10/1:149-55.
1972a Meta-variations, part IV: Analytic fallout (I). *Perspectives of New Music* 11/1:146-223.
1972b Nelson Goodman's *Languages of art* from a musical point of view. In Boretz and Cone (1972a:31-44).
1973 Meta-variations, part IV: Analytic fallout (II). *Perspectives of New Music* 11/2:156-203.
Boretz, Benjamin and Cone, Edward, eds.
1971 *Perspectives on American composers.* New York: W.W. Norton.
1972a *Perspectives on contemporary music theory.* New York: W.W. Norton, 1972.
Boros, James
1990 Some properties of the all-trichordal hexachord. *In Theory Only* 11/6:19-41.
1991 Donald Martino's *Fantasy Variations*: The first three measures. *Perspectives of New Music* 29/2:280-93.
Borthwick, Alastair
1995 *Music theory and analysis; The limitations of logic.* New York: Garland.
Brinkman, Alexander R.
1986 A binomial representation of pitch for computer processing of musical data. *Music Theory Spectrum* 8:44-57.
Brody, Martin
1977 Sensibility defined: Set projection in Stefan Wolpe's *FORM* for piano. *Perspectives of New Music* 15/2:3-22.
1993 Music for the masses: Milton Babbitt's cold war music theory. *Musical Quarterly* 77/2:161-192.
Brouwer, Andries Evert
1989 *Distance-regular graphs.* Berlin: Springer-Verlag.
Browne, Richmond
1974 Review of *The structure of atonal music,* by Allen Forte. *Journal of Music Theory* 18/2:390-409.
1976 The Babbitt-Mallalieu fully cyclically permutational row. *In Theory Only* 2/7:10-14.
1977-8 Index of music theory in the United States:1955-1970. *In Theory Only* 3/7-11.
Brucks, Karen M.
1996 *Organization of parameter space for simple circle maps: The Farey web.* Yorktown Heights, New York: IBM A.J. Watson Research Center.
Buckley, Fred
1990 (with Harary, F.) *Distance in graphs.* Redwood City, CA: Addison-Wesley.
Cameron, Peter
1991 *Designs, graphs, codes, and their links.* Cambridge: Cambridge University Press.

Capalbo, Marc
1981 Charts. *Perspectives of New Music* 19/1-2:309-31.
Capobianco, Michael and Molluzzo, J.C.
1978 *Examples and counterexamples in graph theory.* New York: North-Holland.
Carey, Norman and Clampitt, David
1989 Aspects of well-formed sets. *Music Theory Spectrum* 11/2:187-206.
Carlson, Effie B.
1970 *A bio-bibliographical dictionary of twelve-tone and serial composers.* Metuchen, NJ: Scarecrow Press, (Wolpe, Stefan:187-189).
Carter, Elliott
1960 Shop talk by an American composer. *Musical Quarterly* 46:189-201.
1972 In memoriam [Stefan Wolpe] (1902-1972). *Tempo* 102:187-189.
Carter, Elliott; Babbitt, Milton; Wuorinen, Charles; and Stempel, Larry.
1972 In memoriam: Stefan Wolpe. *Perspectives of New Music* 11/1:3-10.
Cartwright, Dorwin
1966 (with Gleason, A.) The number of paths and cycles in a digraph. *Psychometrika* 31:179-199.
Cayley, Arthur
1895 The theory of groups, graphical representations. *Collected Papers* Cambridge 10:26-28.
1889 A theorem on trees. *Quarterly Journal of Mathematics Oxford Series* 23:376-378; also in *Collected Papers,* Cambridge 13:26-28, 1897.
Chaiken, Seth David
1980 *Matrix tree theorems and degree sequence realization by strongly 2-connected digraphs.* Ph.D. dissertation, M.I.A.: Dept. of Mathematics.
Chao, Chong-Yun
1971 On the classification of symmetric graphs with a prime number of vertices, *Transactions. American Mathematical Society* 158:247-256.
Chapman, Alan
1981 Some intervallic aspects of pitch-class set relations. *Journal of Music Theory* 25/2:275-90.
Chartrand, Gary
1977 *Introductory graph theory.* New York: Dover
1986 (with Lesniak, L.) *Graphs and digraphs.* Monterey: Wadsworth & Brooks/Cole Advanced Books and Software.
Cherlin, Michael
1986 Alphonce's invariance matrix and Lewin's inversional clock: A new approach toward reading pitch-class matrices. *In Theory Only* 9/2-3:47-59.
1988 Review of *Composition with pitch-classes,* by Robert D. Morris. *Intégral* 2:179-200.
Choisnard, Paul
1901 *Influence astrale.* Paris: Chacornac.
1926 *La Chaine harmonies ou la spirale dans la nature.* 2nd ed. Paris: Ernest Leroux.
Chrisman, Richard
1971 Identification and correlation of pitch-sets. *Journal of Music Theory* 15:58-83.
1977 Describing structural aspects of pitch-sets using successive-interval arrays. *Journal of Music Theory* 21/1:1-28.
Clarkson, Austin
1981 *Stefan Wolpe: A brief catalogue of published works.* Islington, Ontario: The Sound Way Press.
1984 Stefan Wolpe's Berlin years. *Music and Civilization: Essays in Honor of Paul Henry Lang*, ed. E. Strainchamps, M. R. Mannniates, C. Hatch. New York: W.W. Norton:371-391.

1985 Spatial proportions, pitch complexes, and organic modes in the later music of Stefan Wolpe (1902-1972). Vancouver. (Lecture.)

Clifton, Thomas

1970 Types of symmetrical relations in Stravinsky's *A sermon, narrative, and a prayer. Perspectives of New Music* 9/1:96-112.

Clough, John

1965 Pitch-set equivalence and inclusion (A comment on Forte's theory of set-complexes). *Journal of Music Theory* 9/1:163-71.

1979 Aspects of diatonic sets. *Journal of Music Theory* 23/1:45-61.

1983 Use of the exclusion relation to profile pitch-class sets. *Journal of Music Theory* 27/2:181-201.

1989 Review of *Generalized musical intervals and transformations,* by David Lewin. *Music Theory Spectrum* 11/2:226-31.

Clough, John and Douthett, Jack

1991 Maximally even sets. *Journal of Music Theory* 35/1:93-174.

Clough, John and Myerson, Gerald

1985 Variety and multiplicity in diatonic systems. *Journal of Music Theory* 29/2:249-70.

Cogan, Robert and Escot, Pozzi

1976 *Sonic design.* Englewood Cliffs, NJ: Prentice-Hall.

Cohn, Richard

1988 Inversional symmetry and transpositional combination in Bartók. *Music Theory Spectrum* 10:19-42.

1989 Review of *Generalized musical intervals and transformations,* by David Lewin. *Journal of Musicological Research* 9/1:47-60.

1991a Properties and generability of transpositionally invariant sets. *Journal of Music Theory* 35/1:1-32.

1991b Bartók's octatonic strategies: A motivic approach. *Journal of the American Musicological Society* 44/2:262-300.

Coons, Edgar and Kraehenbuehl, David

1958 Information as a measure of structure in music. *Journal of Music Theory* 2/2:127-61.

Cope, David

1977 *New music composition.* New York: Schirmer Books.

Coulson, Maria M.

1995 *The mathematical theory of music.* Master's thesis, California State University, Northridge.

Covach, John

1992 The zwölftonspiel of Josef Matthias Hauer. *Journal of Music Theory* 36/1:149-84.

1996 Review of *Music Theory and Analysis; The Limitations of Logic,* by Alastair Borthwick. *Notes* 52/4:1192-94.

Cowell, Henry

1930 *New musical resources.* New York:Knopf.

Dagnes, Edward P.

1976 Symmetrical structures in Webern: An analytical overview of the *Symphonie,* movement 2, variation 3. *In Theory Only* 1:33-51.

Davis, James A.

1995 Philosophical positivism and American atonal music theory. *Journal of the History of Ideas* 56/3:501-22.

DeBruin, Nicholas G.

1946 A combinatorial problem. *Indagationes Mathematicae* 8:461-467.

1964 Polya's theory of counting. in *Applied Combinatorial Mathematics.* Ed. E. F. Beckenbach, New York: Wiley:144-184.

DeLio, Thomas
1980 Iannis Xenakis' *Nomos Alpha*: The dialectics of structure and materials. *Journal of Music Theory* 24/1:63-96.
Dembski, Stephen and Straus, Joseph, ed.
1987 *Milton Babbitt; Words about music.* Madison: The University of Wisconsin Press.
DeYoung, Lynden
1978 Pitch order and duration order in Boulez *Structures IA. Perspectives of New Music* 16/2:27-34.
Diamond, Harold J.
1991 *Music analyses; An annotated guide to the literature.* New York: Schirmer Books.
Dowd, Martin
1982 *Isomorphism of complete sets.* Rutgers University: Laboratory for Computer Science Research.
Dubiel, Joseph
1990 Review of *Composition with pitch-classes,* by Robert D. Morris. *Journal of Musicological Research* 10/1-2:47-80.
Eimert, Herbert
1958 Interval proportions. *Die Reihe* 2:93-99.
Eisinger, Norbert
1991 *Completeness, confluence, and related properties of clause graph resolution.* London: Morgan Kaufmann.
Erdös, Paul
1935 (with Szekeres, G.) A combinatorial problem in geometry. *Composito Mathematica* 2:464-470.
1936 (with Turán, P.) On some sequences of integers. *Journal of the London Mathematical Society* 11:261-264.
1950 (with Rado, R.) A combinatorial theorem. *Journal of the London Mathematical Society* 25:249-255.
1952 (with Rado, R.) Combinatorial theorems on classifications of subsets of a given set. *Proceedings. London Mathematical Society* 2:417-439.
1956 (with Rado, R.) A partition calculus in set theory. *Bulletin of the American Mathematical Society* 62:427-489.
1960 (with Gallai, A.) Graphs with Prescribed Degree of Vertices. [in Hungarian] *Matematikai Lapok* 11:264-274.
1961 Graph theory and probability. *Canadian Journal of Mathematics* 13:346-352.
1963 (with Sachs, H.) Reguläre graphen gegebener taillenweite mit minimaler knotenzahl. *Wissenschaftliche Zeitschrift Martin-Luther-Universitaet (Halle-Wittenberg). Mathematisch-Naturwissenschaftliche Reihe* 12:251-258.
1964 (with Moon, J. W.) On subgraphs of the complete bipartite graphs. *Canadian Mathematics Bulletin* 7:35-39.
1965 (with Harary, F. and Tutte, W.A.) On the dimension of a graph. *Mathematika* 12:118-122.
1968 (with Purdy, George) *Some extremal problems in geometry.* Calgary, Alberta: University of Calgary.
1968 (with Hajnal, A. and Milner, E. C.) A problem on well ordered sets. *Department of Mathematics. Research Paper no. 55.* Calgary: University of Calgary.
1969 (with Sos, V. A.) Some remarks on Ramsey's and Turán's theorems. *Colloquia Mathematica Societatis Janos Bolyai 4. Combinatorial Theory and Its Applications,* Balatonfured, Hungary: North Holland:395-404.
1973 *The art of counting; Selected writings.* Ed. Joel Spencer. Cambridge, MA: MIT Press.

1974 (with Spencer, Joel) *Probabilistic methods in combinatorics.* New York: Academic Press.
1976 (with Burr, S. A.) External Ramsey theory for graphs. *Utilitas Mathematica* 9:247-258.
1980 (with Graham, R.L.) *Old and new problems and results in combinatorial number theory.* Geneve: L' Enseignement Mathematique, Universite de Geneve.
1980 (with Fajtlowicz, S. and Hoffman, A.J.) Maximum degree in graphs of diameter 2. *Networks* 10:87-90.
1984 [et al.] *Combinatorial set theory: Partition Relations for Cardinals.* New York: North-Holland Pub. Co.
1989 (with Gruber, P.M. and Hammer, J.) *Lattice points.* New York: Wiley.
1989 (with Richmond, L.B.) *On graphical partitions.* Waterloo, Ont.: University of Waterloo.
1990 *Collected papers of Paul Turán.* Ed. P. Erdös, Budapest: Akademiai Kiado.
1990 *A tribute to Paul Erdös.* Ed. A. Baker, B. Bollobás, A. Hajnal, Cambridge: Cambridge University Press.

Erickson, Robert
1975 *Sound structure in music.* Berkeley: University of California Press.

Ericksson, Tore
1986 The IC max point structure, MM vectors and regions. *Journal of Music Theory* 30/1:95-111.

Eschman, Karl
1945 *Changing forms in modern music.* Boston: E.C. Schirmer.

Escot, Pozzi
1988 'Charm'd magic casements;' mathematical models in Ligeti. *Sonus* 9/1:17-37.

Euler, Leonhard
1736 Solutio problematis ad geometriam situs pertinentis. *Comment. Academiae Science I. Petropolitanae* 8:128-140. Also in *Opera Omnia Series* I-7 (1766) 1-10.

Ewen, David
1982 *American composers: A biographical dictionary.* New York: G.P. Putnam's Sons, [Wolpe, Stefan:726-729].

Forte, Allen
1955 *Contemporary tone structures.* New York: Teachers College, Columbia University.
1960 Bartók's 'serial' composition. *Musical Quarterly* 46/2:233-45.
1963 Context and continuity in an atonal work. *Perspectives of New Music* 1/2:72-82.
1964 A theory of set-complexes for music. *Perspectives of New Music* 8/2:136-83.
1965 The domain and relations of set-complex theory. *Journal of Music Theory* 9/1:173-80.
1972 Sets and nonsets in Schoenberg's atonal music. *Perspectives of New Music* 11/1:43-64.
1973a *The structure of atonal music.* New Haven: Yale University Press.
1973b The basic interval patterns. *Journal of Music Theory* 17/2:234-73.
1974a Analysis symposium: Webern, *Orchestra pieces* (1913): Movement I (*Bewegt*). *Journal of Music Theory* 18/1:13-43.
1974b Theory (Vinton 1974:753-61).
1978a *The harmonic organization of The Rite of Spring.* New Haven: Yale University Press.
1978b Schoenberg's creative evolution: The path to atonality. *Musical Quarterly* 64/2:133-76.
1980 Aspects of rhythm in Webern's atonal music. *Music Theory Spectrum* 2:90-109.

1981 The magical kaleidoscope: Schoenberg's first atonal masterwork, Op.11, No. 1. *Journal of the Arnold Schoenberg Institute* 5/2:127-68.

1985a Tonality, symbol, and structural levels in Berg's *Wozzeck*. *Musical Quarterly* 71/4:474-99.

1985b Pitch-class set analysis today. *Music Analysis* 4/1-2:29-58.

1990 A major Webern revision and its implications for analysis. *Perspectives of New Music* 28/1:224-55.

1991 The mask of tonality; Alban Berg's symphonic epilogue to *Wozzeck*. *Alban Berg: Historical and Analytical Perspectives.* Morgan, Gable (1991:151-200).

1992 Concepts of linearity in Schoenberg's atonal music: A study of the Opus 15 song cycle. *Journal of Music Theory* 36/2:285-382.

Fotine, Larry

1967 *Theory and techniques of twelve-tone composition.* Sepulveda: Polytone Music Publishing Company.

Friedmann, Michael L.

1985 A methodology for the discussion of contour: Its application to Schoenberg's music. *Journal of Music Theory* 29/2:223-48.

Fuller, Ramon

1972 A study of interval and trichord progressions. *Journal of Music Theory* 16:102-41.

Gamer, Carlton

1967a Deep scales and difference sets in equal-tempered systems. *Proceedings of the American Society of University Composers* 2:113-22.

1967b Some combinational resources of equal tempered systems. *Journal of Music Theory* 11/1:32-59.

1974 Piano raga music. *Perspectives of New Music* 12:191-230.

1981 Et setera: Some temperamental speculations. In *Music Theory: Special Topics.* Ed. Richmond Browne, New York: Academic Press.

Gauldin, Robert

1977a The magic squares of the third movement of Webern's *Concerto,* Op. 24. *In Theory Only* 2/9:32-42.

1977b Pitch structures in the second movement of Webern's *Concerto,* Op. 24. *In Theory Only* 2/10:8-22.

1978 A pedagogical introduction to set-theory. *Theory and Practice* 3/2:3-14.

1983 The cycle-7 complex; Relations of diatonic set theory to the evolution of ancient tonal systems. *Music Theory Spectrum* 5:39-55.

1985 (with Benson, Warren) Structure and numerology in Stravinsky's *In memorium Dylan Thomas*. *Perspectives of New Music* 23/2: 166-85.

1988 Review of *The music of Edgard Varèse,* by Jonathan Bernard and *The Music of Paul Hindemith,* by David Neumeyer. *Music Theory Spectrum* 10:137-42.

Gibbons, Alan

1985 *Algorithmic graph theory.* Cambridge: Cambridge University Press.

Gilbert, Edgar N.

1956 Enumeration of labeled graphs. *Canadian Journal of Mathematics* 8:405-411.

Gilbert, Steven

1970 The twelve-tone system of Carl Ruggles: A study of the *Evocations* for piano. *Journal of Music Theory* 14/1:68-91.

1974 An introduction to trichordal analysis. *Journal of Music Theory* 18/2:338-62.

Gilmore, Bob

1995 Changing the metaphor; Ratio models of musical pitch in the work of Harry Partch, Ben Johnston, and James Tenny. *Perspectives of New Music* 33/1-2:458-503.

Goder, Miriam

1972 The interval triangle. *Journal of Music Theory* 16/1:142-67.

Godwin, Joscelyn
1995 *Music and the occult; French musical philosophies 1750-1950.* Rochester: University of Rochester Press.
Goodman, Nelson
1966 *The structure of appearance.* Indianapolis: Bobbs-Merrill.
1976 *Languages of art: An approach to a theory of symbols.* Indianapolis: Hackett.
Gustafson, Roger
1977 *The theories of Josef Matthias Hauer: An English translation of selected writings with critical commentary and a list of works.* Ph.D. dissertation, Michigan State University.
Hage, Per
1991 (with Harary, Frank) *Exchange in Oceania; A graph-theoretical analysis.* Oxford: Clarendon Press.
Haimo, Ethan
1984 (with Johnson, Paul) Isomorphic partitioning and Schoenberg's *Fourth String Quartet. Journal of Music Theory* 28/1:47-72.
Hanson, Howard
1960 *The harmonic materials of twentieth-century music.* New York: Appleton-Century Crofts.
Harary, Frank
1955 The number of linear, directed, rooted, and connected graphs. *Transactions. American Mathematical Society* 78:445-463.
1958 Note on Carnap's relational asymptotic relative frequencies. *Journal of Symbolic Logic* 23:257-260.
1959 (with Prins, G.). The number of homeomorphically irreducible trees, and other species. *Acta Mathematica* 101:141-162.
1962 The determinant of the adjacency matrix of a graph. *SIAM Review* 4:202-210.
1964 (with Prins, G. and Tutte, W. A.). The number of plane trees. *Indagationes Mathematicae* 26:319:329.
1965 (with Norman, Robert Z. and Cartwright, Dorwin) *Structural models: An introduction to the theory of directed graphs.* New York: Wiley.
1967 (with Palmer, E. M. and Read, R. C.). The number of ways to label a structure. *Psychometrika* 32:155-156.
1967 Applications of Polya's theorem to permutation groups. in *A Seminar on Graph Theory.* Ed. F. Harary, New York: Holt:25-33.
1967 *Graph theory and theoretical physics.* New York: Academic Press.
1967 Graphical enumeration problems. in *Graph Theory and Theoretical Physics.* Ed. F. Harary, New York: Academic Press:1-41.
1967 (Ed. with Beineke, L.) *A seminar on graph theory.* New York: Holt, Rinehart and Winston.
1968 *Determinants, permanents, and bipartite graphs.* Santa Monica, CA: Rand Corp.
1969 *Graph theory.* Reading: Addison-Wesley.
1969 (with Mowshowitz, A. and Riordan, J.). Labeled trees with unlabeled end-points. *Journal of Combinatorial Theory* 6:60-64.
1971 (with Manvel, B.). On the number of cycles in a graph. *Matematicky Casopis Slovenskej Akademie Vied* 21:63.
1973 *New directions in the theory of graphs; Proceedings.* Ed. F. Harary, New York: Academic Press.
1973 (with Palmer, Edgar M.) *Graphical enumeration.* New York:Academic Press.
1985 (Ed. with Maybee, John) *Graphs and applications; Proceedings of the first Colorado Symposium on Graph Theory.* New York: Wiley.
Hartsfield, Nora and Ringel, Gerhard
1985 (with Jackson, B) The splitting number of the complete graph. *Graphs and Combinatorics* 1: 311-329.

1990 *Pearls in graph theory; A comprehensive introduction.* New York: Academic Press.

Harvey, Dixie Lynn

1980 *The theoretical treatises of Josef Matthias Hauer.* Ph.D. dissertation, North Texas State University.

Hasty, Christopher

1978 A theory of segmentation developed from late works of Stefan Wolpe. Ph.D. dissertation, Yale University.

1981a Segmentation and process in post-tonal music. *Music Theory Spectrum* 3:54-73.

1981b Rhythm in post-tonal music; Preliminary questions of duration and motion. *Journal of Music Theory* 25/2:183-216.

1984 Phrase formation in post-tonal music. *Journal of Music Theory* 28/2:167-190.

1986 On the problem of succession and continuity in twentieth-century music. *Music Theory Spectrum* 8:58-74.

Hauer, Josef Matthias

1920 Musikalische bildung. *Melos* 1:458-9.

1922 Melos und rhythmus. *Melos* 3:186-7.

1922 Sphärenmusik. *Melos* 3:132-3.

1923 *Deutung des melos: Eine frage an die künstler und denker unserer zeit.* Vienna: E.P. Tal and Co.

1923 *Vom wesen des musikalischen: Grundlagen der zwölftonmusik.* Vienna: Schlesinger. Ed. Victor Sokolowski. Berlin: Robert Lienau. The reprinted edition contains additional trope charts from the Hauer estate Reprinted:1948.

1923 Atonale musik *Die Musik* 16:103.

1925 *Vom melos zur pauke: Eine einfuhrung in die zwölftonmusik.* Vienna.

1926 *Zwölftontechnik: die lehre von den tropen.* Vienna: Schlesinger.

Headlam, Dave

1985 The derivation of rows in *Lulu. Perspectives of New Music* 24/1:198-233.

1990 Row derivation and contour association in Berg's *Der Wein. Perspectives of New Music* 28/1:256-93.

Helin, Jacquelyn Mae

1982 'The ever-restored and ever-advancing moment': Stefan Wolpe's compositional philosophy and aesthetic as seen in 'Form' for piano. D.M.A. dissertation, The University of Texas at Austin.

Ho, Yong Song

1969 *Theory of graphs; a bibliography.* Singapore: Lee Kong Chian Institute of Mathematics, Nanyang University.

Hofstadter, Douglas

1979 *Gödel, Escher, Bach: An eternal golden braid.* New York: Basic Books.

Holton, Derek

1993 *The Petersen graph.* Cambridge: Cambridge University Press.

Hoover, Mark

1984 Set constellations. *Perspectives of New Music* 23/1:164-79.

Howe, Hubert S., Jr.

1965 Some combinational properties of pitch structures. *Perspectives of New Music* 4/1:45-61.

Hyatt, Allyson Leigh

1996 *Hamiltonian problems in graph theory.* Master's thesis, Wake Forest University.

Isaacson, Eric J.

1990 Similarity of interval-class content between pitch-class sets: The IcVSIM relation. *Journal of Music Theory* 34/1:1-28.

Jackendoff, Ray and Lerdahl, Fred
1980 Discovery procedures vs. rules of musical grammar in a generative music theory. *Perspectives of New Music* 18/1:503-10.
Jackson, Bradley and Ringel, Gerhard
1984 The splitting number of complete bipartite graphs. *Archiv der Mathematik* 42: 178-184.
Johnson, Jeffrey
1995 *Thesaurus of abstract musical properties.* Westport, CT: Greenwood Press.
Kabbash, Paul
1984 Aggregate-derived symmetry in Webern's early works. *Journal of Music Theory* 28/2:225-50.
Kagel, Mauricio
1965 Translation-rotation. *Die Reihe* 7:32-60.
Kaplan, Richard A.
1990 Transpositionally invariant subsets: A new set-subcomplex relation. *Intégral* 4:37-66.
Kassler, Michael
1963 A sketch of the use of formalized languages for the assertion of music. *Perspectives of New Music* 1/2:83-94.
1967 Toward a theory that is the twelve-note-class system. *Perspectives of New Music* 5/2:1-80. Also in Kassler (1970).
Katz, Adele
1946 *Challenge to musical tradition; A new concept of tonality.* New York: Alfred A. Knopf.
Kendall, Gary
1981 Composing from a geometric model: Five-leaf rose. *Computer Music Journal* 5/5:66-73.
Kerr, Elizabeth
1984 The variations of Webern's *Symphony* Op. 21: Some observations on rhythmic organization and the use of numerology. *In Theory Only* 8/2:5-14.
Klammer, Armin
1958 Webern's *Piano Variations,* Op. 27, 3rd Movement. *Die Reihe* 2:81-92.
Kostka, Stefan
1990 *Materials and techniques of twentieth-century music.* Englewood Cliffs, NJ: Prentice-Hall.
Kowalski, David
1986 An annotated bibliography of array studies. *Indiana Theory Review* 7/1:52-65.
Krenek, Ernst
1960 Extents and limits of serial techniques. *Musical Quarterly* 46:210-32.
Kyaw, Shwe
1994 *A dirac-type criterion for Hamiltonicity.* Berlin: Verlag Koster.
Lake, William
1986 The architecture of a superarray composition: Milton Babbitt's *String Quartet No. 5. Perspectives of New Music* 24/2:88-111.
Lambert, J. Phillip
1990 Interval-cycles as compositional resources in the music of Charles Ives. *Music Theory Spectrum* 12/1:43-82.
Lansky, Paul
1973 *Affine music.* Ph.D. dissertation, Princeton University. See Perle (1977).
1975 Pitch-class consciousness. *Perspectives of New Music* 13/2:30:56.
Lendvai, Ernö
1971 *Béla Bartók: An analysis of his music.* London: Kahn and Averill.

Levarie, Siegmund and Levy, Ernst
1983 *Musical morphology: A discourse and a dictionary.* Kent, OH: Kent State University Press.

Levy, Edward
1967 A guide to musical analysis. Ed.D. dissertation, Columbia University [Stefan Wolpe:379-407].

Lewin, David
1959 Intervallic relations between two collections of notes. *Journal of Music Theory* 3/2:298-301.
1960 The intervallic content of a collection of notes, intervallic relations between a collection of notes and its complement: An application to Schoenberg's hexachordal pieces. *Journal of Music Theory* 4/1:98-101.
1962a A metrical problem in Webern's Op. 27. *Journal of Music Theory* 6/1:124-32.
1962b A theory of segmental association in twelve-tone music. *Perspectives of New Music* 1/1:89-116. Also in Boretz and Cone (1972a:180-207).
1966 On certain techniques of re-ordering in serial music. *Journal of Music Theory* 10/2:276-87.
1967a *Moses and Aron*: Some general remarks, and analytical notes for Act 1, Scene 1. *Perspectives of New Music* 6/1:1-17. Also in Boretz and Cone (1972b).
1967b A study of hexachordal levels in Schoenberg's *Violin Fantasy. Perspectives of New Music* 6/1:18-32. Also in Boretz and Cone (1972b).
1968a Inversional balance as an organizing force in Schoenberg's music and thought. *Perspectives of New Music* 6/2:1-21.
1968b Some applications of communication theory to the study of twelve-tone music. *Journal of Music Theory* 12/1:50-84.
1969 Behind the beyond: A response to Edward A. Cone. *Perspectives of New Music* 7/2:59-69. See Cone (1967).
1974 Toward the analysis of a Schoenberg song (Op. 15, No. 11). *Perspectives of New Music* 12:43-86.
1976a On partial ordering. *Perspectives of New Music* 15/1:252-59.
1976b On the interval content of invertible hexachords. *Journal of Music Theory* 20/2:185-8.
1976c Letter (On Mallalieu row). *In Theory Only* 2/7:8.
1977a Forte's interval vector, my interval function, and Regener's common-note function. *Journal of Music Theory* 21/2:194-237.
1977b A label-free development of 12-pitch-class systems. *Journal of Music Theory* 21/1:29-48.
1977c Some notes on Schoenberg's Op. 11. *In Theory Only* 3/1:3-7.
1980a Some new constructs involving abstract PC sets, and probabilistic applications. *Perspectives of New Music* 18/2:433-44.
1980b A response to a response: On PC set relatedness. *Perspectives of New Music* 18/1:498-502.
1981 A way into Schoenberg's Op. 15, No. 7. *In Theory Only* 6/1:3-24.
1982 Transformational techniques in atonal and other music theories. *Perspectives of New Music* 21/1-2:312-71
1982 A formal theory of generalized tonal functions. *Journal of Music Theory* 26/1:23-60.
1984 On formal interval between time-spans. *Music Perception* 1/4:414-23.
1986 Music theory, phenomenology, and modes of perception. *Music Perception* 3/4: 327-92.
1987 *Generalized musical intervals and transformations.* New Haven: Yale University Press.

1990 Klumpenhouwer networks and some isographies that involve them. *Music Theory Spectrum* 12:83-120.
1991 Some problems and resources of music theory. *Journal of Music Theory Pedagogy* 5/2:111-32.
1993 *Musical form and transformation: Four analytic essays.* New Haven: Yale University Press.

Lewin, David; Babbitt, Milton; Browne, Richmond; and Morris, Robert
1976 Maximally scrambled 12-tone sets: A serial forum. *In Theory Only* 2/5:13-26.

Lewin, Harold F.
1965 *Aspects of the twelve-tone system: its formation and structural implications.* Ph.D. dissertation, Indiana University.
1978 Some basic concepts for determining the combinational properties of 12-note set segments (I). *Theory and Practice* 3/1:17-28.

Liang, Y. Daniel
1993 *NCp2s algorithms regarding Hamiltonian paths and circuits in interval graphs.* Durham, NH: Dept. of Computer Sciences, University of New Hampshire.

Lindley, Mark
1993 *Mathematical models of musical scales: A new approach.* Bonn:Verlag für Systematische Musikwissenschaft.

Lord, Charles
1981 Intervallic similarity relations in atonal set analysis. *Journal of Music Theory* 25/1:91-111.

Lu, Xiaoyun
1992 *Hamilton cycles and games on graphs.* DIMACS: Center for Discrete Mathematics and Theoretical Computer Science.

Martino, Donald
1961 The source set and its aggregate formations. *Journal of Music Theory* 5:224-73. (Addendum in *Journal of Music Theory*:322-3.)

Marvin, Elizabeth West
1983 The structural role of complementation in Webern's *Orchestra Pieces (1913). Music Theory Spectrum* 5:76-88.

Mead, Andrew
1981 *Set structure and extended form.* Ph.D. dissertation, Princeton University.
1983 Detail and the array in Milton Babbitt's *My Complements to Roger. Music Theory Spectrum* 5:89-109.
1983-4 Pitch structure in Elliott Carter's *String Quartet No. 3. Perspectives of New Music* 22/1 and 22/2:31.
1984 Letter to the editor: Reply. *In Theory Only* 8/3:4-5. See Lake (1984).
1985 Large-scale strategy in Arnold Schoenberg's twelve-tone music. *Perspectives of New Music* 24/1:120-157.
1987 About *About Times*'s time: A survey of Milton Babbitt's recent rhythmic practice. *Perspectives of New Music* 25:182-235.
1988 Some implications of the pitch-class/order number isomorphism inherent in the twelve-tone system: Part one. *Perspectives of New Music* 26/2:96-163.
1989a Some implications of the pitch-class/order number isomorphism inherent in the twelve-tone system: Part two: The Mallalieu complex, its extensions and related rows. *Perspectives of New Music* 27/1:180-223.
1989b Twelve-tone organizational strategies: An analytical sampler. *Intégral* 3:93-170.
1989c The state of research in twelve-tone and atonal theory. *Music Theory Spectrum* 11/1:40-49.
1991 Review of *Composition with Pitch-Classes,* by Robert D. Morris. *Perspectives of New Music* 29/1:264-311.
1992 Review of *The Twelve-Note Music of Anton Webern,* by Kathryn Bailey. *Intégral* 6:107-35.

1993 Webern, tradition, and 'composing with twelve tones.' *Music Theory Spectrum* 15/2:173-204.

1993 *An introduction to the music of Milton Babbitt.* Princeton: Princeton University Press.

Messiaen, Olivier

1944 *Technique de mon language musical.* Paris: Leduc.

Meyer, Leonard B.

1956 *Emotion and meaning in music.* Chicago: University of Chicago Press.

Michael, A. S.

1988 *The structure matrix of the class of R-multigraphs with a prescribed degree sequence.* Ph.D. dissertation, University of Wisconsin.

Milstein, Silvina

1992 *Arnold Schoenberg: Notes, sets, forms.* Cambridge: Cambridge University Press.

Moekle, Janet Wickline

1986 Schillinger's notation; Review of *Graph method of dance notation* by Joseph Schillinger. *Dance Chronicle* 9/3:403-4.

Mole, Richard H.

1989 *BASIC graph and network algorithms.* London: Butterworths.

Molloy, Michael, O'Brian, Sean

1994 *Random graphs with a fixed degree sequence.* Ph.D. dissertation, Carnegie Mellon University.

Monelle, Raymond

1996 Review of *A theory of musical seminotics* by Eero Tarasti. *Music and Letters* 77/1:147-9.

Moon, John

1967 Various proofs of Cayley's formula for counting trees. in *A Seminar on Graph Theory.* Ed. F. Harary, New York: Holt:70- 78.

1970 *Counting labeled trees.* Montreal: Canadian Mathematical Congress.

Morgan, Robert

1975 Stockhausen's writings on music. *Musical Quarterly* 61/1:1-16.

1977 On the analysis of recent music. *Critical Inquiry* 4/1:33-53.

1991a *Twentieth-century music: A history of musical style in modern Europe and America.* New York: W.W. Norton.

1991b The eternal return; retrograde and circular form in Berg. In *Alban Berg: Historical and analytical perspectives.* Morgan, Gable (1991:111-149).

Morris, Robert D.

1974 (with Starr, Daniel) The structure of the all-interval series. *Journal of Music Theory* 18/2:364-89.

1976 More on 0, 1, 4, 2, 9, 5, 11, 3, 8, 10, 7, 6. *In Theory Only* 2/7:15-20.

1977 (with Starr, Daniel) A general theory of combinatoriality and the aggregate. *Perspectives of New Music* 16/1:364-89 and 16/2:50-84.

1977 On the generation of multiple order-function twelve-tone rows. *Journal of Music Theory* 21/2:238-63.

1980 A similarity index for pitch-class sets. *Perspectives of New Music* 18/2:445-60.

1982a Review of *Basic atonal theory,* by John Rahn. *Music Theory Spectrum* 4:138-54.

1982b Set groups, complementation, and mappings among pitch-class sets. *Journal of Music Theory* 26/1:101-44.

1983a Combinatoriality without the aggregate. *Perspectives of New Music* 21/1-2:432-86.

1983b Set-type saturation among twelve-tone rows. *Perspectives of New Music* 22/1:187-217.

1985 Review of *The music of Elliott Carter,* by David Schiff. *Journal of the American Musicological Society* 38:183-90.
1987 *Composition with pitch-classes: A theory of compositional design.* New Haven and London: Yale University Press.
1988 Generalizing rotational arrays. *Journal of Music Theory* 32/1:75-132.
1988 (with Alegant, Brian) The even partitions in twelve-tone music. *Music Theory Spectrum* 10:74-101.
1989 Some compositional and analytical applications of T-matrices. *Intégrall* 3:37-66. Errata listed in *Intégral* 4:117-18.
1990 Pitch-class complementation and its generalizations. *Journal of Music Theory* 34/2:175-246.
1991 *Class notes for atonal music theory.* Lebanon, NH: Frog Peak Music.
1993 New directions in the theory and analysis of musical contour. *Music Theory Spectrum* 15/2:205-228.
1995a Equivalence and similarity in pitch and their interaction with pcset theory. *Journal of Music Theory* 39/2:207-244.
1995b Review of *Musical form and transformation; Four analytic essays,* by David Lewin. *Journal of Music Theory* 39/2:342-383.
1995c Compositional spaces and other territories. *Perspectives of New Music* 33/1-2:328-359.

Muzzulini, Daniel
1995 Musical modulation by symmetries. *Journal of Music Theory* 39/2:311-328.

Nauert, Paul
1994 A theory of complexity to constrain the approximation of arbitrary sequences of timepoints. *Perspectives of New Music* 32/2:226-263.

Otter, R.
1948 The number of trees. *Annals of Mathematics* 49:583-599.

Palmer, Edgar M.
1969 On the number of labeled 2-trees. *Journal of Combinatorial Theory* 6:206-207.
1969 Combinatorial problems in set-theoretic form. *Mathematics Magazine.* 42:32-37.
1969 Methods for the enumeration of multigraphs. in *The Many Facets of Graph Theory.* Ed. G. Chartrand and S. Kapoor. New York: Springer Publ.:251-261.
1970 Asymptotic formulas for the number of self-complementary graphs and digraphs. *Mathematika* 17:85-90.
1972 Graphical enumeration methods. in *New Directions in the Theory of Graphs.* Ed. F. Harary, New York: Academic Press,
1967 (with Robinson, R. W.). The matrix group of two-permutation groups. *Bulletin of the American Mathematical Society.* 73:204-207.

Parncutt, Richard
1994 Applying psychoacoustics in composition; 'harmonic' progressions of 'nonharmonic' sonorities. *Perspectives of New Music* 32/2:88-129.

Parthasarathy, Kalyanapuram
1968 Enumeration of ordinary graphs with given partition. *Canadian Journal of Mathematics* 20:40-47.

Peel, John
1976 On some celebrated measures of the Schoenberg *String Trio. Perspectives of New Music* 14/2–15/1:260-79.
1989 Review of *Composition with pitch-classes,* by Robert D. Morris. *Journal of Music Theory* 33/2:400-416.

Peles, Stephen
1984 Interpretations of sets in multiple dimensions: Notes on the second movement of Arnold Schoenberg's *String Quartet #3. Perspectives of New Music* 22/1-2:303-52.

Perle, George

1954 The possible chords in twelve-tone music. *Score* 9:54-58.

1957 Review of *The hexachord and its relation to the twelve-tone row,* by George Rochberg. *Journal of the American Musicological Society* 10/1:55-9.

1959 The music of *Lulu*: A new analysis. *Journal of the American Musicological Society* 12/2 and 12/3:185-200.

1962 *Serial composition and atonality: An introduction to the music of Schoenberg, Berg and Webern.* Berkeley: University of California Press.

1963 Babbitt, Lewin, and Schoenberg: A critique. *Perspectives of New Music* 2/1:120-7. See also Babbitt (1963) and Lewin (1963).

1964a An approach to simultaneity in twelve-tone music. *Perspectives of New Music* 3/1:91-101.

1964b *Lulu*: The formal design. *Journal of the American Musicological Society* 17/2:179-92.

1965 *Pierrot Lunaire.* In *The Commonwealth of Music.* Ed. Gustave Reese and Rose Brandel. New York: Free Press:307-12.

1967 The musical language of *Wozzeck. Music Forum* 1:204-59.

1971 Webern's twelve-tone sketches. *Musical Quarterly* 57/1:1-25.

1977a *Twelve-tone tonality.* Berkeley: University of California Press.

1977b Berg's master array of interval cycles. *Musical Quarterly* 63/1:1-30.

1980 *The operas of Alban Berg volume I: 'Wozzeck.'* Berkeley: University of California Press.

1984 Scriabin's self-analyses. *Music Analysis* 3/2:101-22.

1985 *The operas of Alban Berg volume II: 'Lulu.'* Berkeley: University of California Press.

1990a *The listening composer.* Berkeley: University of California Press.

1990b Pitch-class set analysis: An evaluation. *The Journal of Musicology* 8:101.

Petersen, J.

1891 Die theorie der regularen graphen. *Acta Mathematica* 15:193-220.

Pope, Conrad

1980 Review of *Simple composition,* by Charles Wuorinen. *Journal of Music Theory* 24/2:253-68.

Pople, Anthony, ed.

1994 *Theory, analysis, and meaning in music.* Cambridge: Cambridge University Press.

Polya, George

1937 Kombinatorische anzahlbestimmungen für gruppen, graphen und chemische verbindungen. *Acta Math.* 68:145-254.

1940 Sur les types des propositions composees. *Journal of Symbolic Logic* 5:98-103.

Post, Nora

1981 Varèse, Wolpe and the oboe. *Perspectives of New Music* 20/1-2:35-148.

Pousseur, Henri

1959 Outline of a method. *Die Reihe* 3:44-88.

1966 The question of order in new music. *Perspectives of New Music* 5/1:93-111. Also in Boretz and Cone (1972a:97-115).

1972 Stravinsky by way of Webern: The consistency of a syntax. *Perspectives of New Music* 10/2:13-51 and 11/1:112-45.

Quine, Willard Van Orman

1960 *Word and object.* Cambridge: Harvard University Press.

1969 *Set theory and its logic.* Rev. Ed. Cambridge: Harvard University Press.

1980 *Elementary logic.* Rev. Ed. Cambridge: Harvard University Press.

1982 *Methods of logic.* Fourth ed. Cambridge: Harvard University Press.

Rahn, Jay

1983 *A theory for all music.* Toronto: University of Toronto Press.

1991 Coordination of interval-sizes in seven-tone collections. *Journal of Music Theory* 35/1:33-60.
1992 An advance on *A theory for all music*: at-least-as predicates for pitch, time, and loudness. *Perspectives of New Music* 30/1:158-183.

Rahn, John
1975a On pitch and rhythm: Interpretation of orderings of and in pitch and time. *Perspectives of New Music* 13/2:182-203.
1975b Gentle reminder #1: Two common-tone theorems. *In Theory Only* 1/2:10-11.
1976 How do you *Du* (by Milton Babbitt)? *Perspectives of New Music* 15/1:61-80.
1979 Aspects of musical explanation. *Perspectives of New Music* 17/2:204-24.
1980a *Basic atonal theory*. New York: Longman.
1980b Relating sets. *Perspectives of New Music* 18/2:483-98.
1987 Review of *Generalized musical intervals and transformations*, by David Lewin. *Journal of Music Theory* 31/2:305-17.

Raines, Terry
1975 An inconsistency in Kassler's first twelve-note-class system. *Perspectives of New Music* 14/1:209-12.

Ramsey, Frank P.
1930 On a problem of formal logic. *Proceedings of the London Mathematics Society* 30:246-286.

Regener, Eric
1974 On Allen Forte's theory of chords. *Perspectives of New Music* 13/1:191-212.

Reid, John W.
1974 Properties of the set explored in Webern's *Variations* Op. 30. *Perspectives of New Music* 12:344-50.

Reti, Rudolph
1958 *Tonality, atonality, pantonality: A study of some trends in twentieth-century music*. London: Barrie and Rockliff.

Riordan, J.
1958 *An introduction to combinatorial analysis*. New York: Wiley.
1960 The enumeration of trees by height and diameter. *IBM Journal of Research and Development* 4:473-78.

Rochberg, George
1955 *The hexachord and its relation to the twelve tone row*. Bryn Mawr: Presser.
1959 The harmonic tendency of the hexachord. *Journal of Music Theory* 3/2:208-30.
1962 Webern's search for harmonic identity. *Journal of Music Theory* 6/1:109-22.

Roeder, John
1987 A geometric representation of pitch-class series. *Perspectives of New Music* 25:362-409.
1989 Review of *Composition with pitch-classes*, by Robert D. Morris. *Music Theory Spectrum* 11/2:240-51.

Rogers, John
1967 Toward a system of rotational arrays. *Proceedings of the American Society of University Composers* 2:61-84.
1968a Some properties of non-duplicating rotational arrays. *Perspectives of New Music* 7/1:80-102.
1968b Some experiments with contextually defined 'octaves.' *Proceedings of the American Society of University Composers* 3:109-22.

Rothgeb, John
1966 Some uses of mathematical concepts in theories of music. *Journal of Music Theory* 10/2:200-215.
1967 Some ordering relationships in the twelve-tone system. *Journal of Music Theory* 11/2:176-97.

Rothstein, William
1980 Linear structure in the twelve-tone system: An analysis of Donald Martino's *Pianississimo*. *Journal of Music Theory* 24/2:129-65.
Rufer, Joseph
1969 *Composition with twelve notes related only to one another*. Trans. Humphrey Searle. London: The Cresset Press.
Russell, Roberta Cram
1983 *A set of microcomputer programs to aid in the analysis of atonal music*. D.M.A. dissertation, University of Oregon.
Salzer, Felix
1952 *Structural hearing*. New York: Dover Publications, reprinted:1982.
Samet, Bruce
1987 *Hearing aggregates*. Ph.D. dissertation, Princeton University.
Schillinger, Joseph
1941 *The Schillinger system of musical composition*. New York: Da Capo Press, reprinted:1978.
ca.1942 *Graph method of dance notation*. Typescript with emendations in ink. New York Public Library: *MGRN-Res. (see also Moekle 1986).
1943 *Papers, ca.1915-1943*. Music Division, The New York Public Library for the performing Arts.
1948 *The mathematical basis of the arts*. New York: Da Capo Press, reprinted:1966.
Schultz, Donna
1979 *Set theory and its application to compositions by five twentieth-century composers*. Ph.D. dissertation, Michigan State University.
Schuster-Craig, John
1990 An eighth mode of limited transposition. *Music Review* 51/4:296-306.
Schwartz, Elliott and Childs, Barney
1967 *Contemporary Composers on Contemporary Music*. New York: Holt, Rinehart & Winston.
Shifrin, Seymour
1975 Review of *Style and idea*, by Arnold Schoenberg. *Perspectives of New Music* 14/1:174-81.
Shrikhande, Mohan
1991 *Quasi-symmetric designs*. Cambridge: Cambridge University Press.
Smyth, David H.
1984 Review of *Basic atonal theory*, by John Rahn. *Perspectives of New Music* 22/1-2:549-55.
Solomon, Larry
1973 New symmetric transformations. *Perspectives of New Music* 3/2:104-26.
1982 The list of chords, their properties, and use in analysis. *Interface* 11:61-107.
Soderberg, Stephen
1995 Z-related sets as dual inversions. *Journal of Music Theory* 39/1:77-100.
Stadelman, Jeffrey
1991 A symmetry of thought. *Perspectives of New Music* 29/2:402-39.
Stanfield, Michael
1984 Some exchange operations in twelve-tone theory: Part one. *Perspectives of New Music* 23/1:258-77.
Starr, Daniel
1978 Sets, invariance and partitions. *Journal of Music Theory* 22/1:1-42.
1984 Derivation and polyphony. *Perspectives of New Music* 23/1:180-257.
Starr, Mark
1970 Colloquy and review: Webern's palindrome. *Perspectives of New Music* 8/2:127-42.

Stedman, Geoffrey
1990 *Diagram techniques in group theory*. Cambridge: Cambridge University Press.

Stewart, B. M.
1967 Supermagic complete graphs. *Canadian Journal of Mathematics* 19:427-438.

Straus, Joseph
1982a Stravinsky's tonal axis. *Journal of Music Theory* 26/2:261-90.
1982b A principle of voice-leading in the music of Stravinsky. *Music Theory Spectrum* 4:106-24.
1984 *Tristan* and Berg's *Lyric suite*. *In Theory Only* 8/3:33-41.
1986 Recompositions by Schoenberg, Stravinsky, and Webern. *Musical Quarterly* 72/3:301-28.
1990 *Introduction to post-tonal theory*. Englewood Cliffs, NJ: Prentice-Hall.

Swift, Richard
1976 Some aspects of aggregate composition. *Perspectives of New Music* 15/1:236-48.
1982 A tonal analog: The tone-centered music of George Perle. *Perspectives of New Music* 21/1:257-85.

Takemitsu Toru
1984 *Dream and number; The mode of musical language*. Tokyo: Studio 200.

Tamassia, Roberto and Eades, Peter
1989 *Algorithms for drawing graphs : An annotated bibliography*. Providence, RI: Brown University.

Tarasti, Eero
1994 *A theory of musical semiotics*. Bloomington: Indiana University Press.

Teck, Katherine
1993 *Ear-training for the body; A dancer's guide to music*. Pennington, NJ: Princeton Book Company.

Teitelbaum, Richard
1965 Intervallic relations in atonal music. *Journal of Music Theory* 9/1:72-127.

Toschi, Andrea
1985 Un modello geometrico del sistema tonale. *Nuova rivista musicale Italiana, Trimestrale di cultura e informazione musicale* 19:458-70.

Tremblay, George
1974 *The definitive cycle of the twelve-tone row*. New York: Criterion Music Corp.

Trinajstic, Nenad
1992 *Chemical graph theory*. Boca Raton, FL: CRC Press.

Trotter, William, ed.
1993 *Planar graphs*. DIMACS vol.9. Providence: American Mathematical Society.

Trudeau, Richard J.
1973 *Introduction to graph theory*. Kent, OH: Kent State University Press, reprinted New York: Dover, 1993.

Vander Weg, John
1979 An annotated bibliography of articles on serialism 1955-1980. *In Theory Only* 5/1:1-36.
1983 *Symmetrical pitch and equivalence class set structure in Anton Webern's Op. 5*. Ph.D. dissertation, University of Michigan.

Verrall, John
1962 A method for finding symmetrical hexachords in serial form. *Journal of Music Theory* 6/2:277-82.

Vinton, John, ed.
1974 *Dictionary of contemporary music*. New York: E.P. Dutton.

Vries, Leo de
1984 Symmetrische diagrammen in de löwintencirkel en de cirkels van de kwarttoonsmuziek. *Mens En Melodie* 39:532-6.

Vuza, Dan Tudor
1988 Some mathematical aspects of David Lewin's book *Generalized musical intervals and transformations*. *Perspectives of New Music* 26:258-87.
1991 Supplementary sets and regular complementary unending canons. *Perspectives of New Music* 29:22-49; 30:184-207; 30:102-25; 31:270-305.

Watkins, Glenn
1988 *Soundings: Music in the twentieth century*. New York: Schirmer Books.

Weiss, Randell Brent
1979 *Finding isomorph classes for combinatorial structures*. Master's thesis, MIT: Dept. of Electrical Engineering and Computer Science.

Wennerstrom, Mary H.
1990 Kostka, Lester, and Straus; Twentieth-century materials and analytical techniques in review. *Intégral* 4:105-16.

Werts, Daniel
1983 *A theory of scale references*. Ph.D. dissertation, Princeton University.

White, Arthur
1973 *Graphs, groups and surfaces*. Amsterdam: North-Holland Publishers.

Whittall, Arnold
1967 Post-twelve-tone analysis. *Proceedings of the Royal Musical Association* 94:12-16.
1977 *Music since the first world war*. London: J.M. Dent.
1989 Review of *Composition with pitch-classes*, by Robert Morris. *Music and Letters* 70/1:134-6.
1991 Review of Kostka (1990), Lester (1989), Simms (1986), Straus (1990), and Watkins (1988). *Music Theory Spectrum* 13/1:79-88.

Wilcox, Howard
1983 Group tables and the generalized hexachord theorem. *Perspectives of New Music* 21/2:535-39.

Wilcox, Howard and Pozzi, Escot
1979 A musical set theory. *Theory and Practice* 4/2:17-37.

Wilson, Richard
1979 *Introduction to graph theory*. New York: Academic Press.
1992 (with van Lint, J. H.) *A course in combinatorics*. Cambridge: Cambridge University Press.

Winham, Godfrey
1970 Composition with arrays. *Perspectives of New Music* 9/1:43-67.

Winick, Steven
1986 Symmetry and pitch-duration associations in Boulez's *Le marteau sans maitre*. *Perspectives of New Music* 24/2:280-321.

Wintle, Christopher
1974 Multiple order functions in twelve-tone music: An informal addendum. *Perspectives of New Music* 12:386-89.
1975 An early version of derivation; Webern's Op. 11, No. 3. *Perspectives of New Music* 13/2:166-77.
1976 Milton Babbitt's *Semi-simple variations Perspectives of New Music* 15/1:111-54.
1980 Schoenberg's harmony: theory and practice. *Journal of the Arnold Schoenberg Institute* 1:50-67.
1981 Analysis and performance; Webern's *Concerto* Op. 24, No. 2. *Music Analysis* 1/1:73-99.

Wittlich, Gary E.
1974 Interval set structure in Schoenberg's Op.11, No. 1. *Perspectives of New Music* 13/1:41-55.

1975 Sets and ordering procedures in twentieth century music. In *Aspects of 20th Century Music*. Ed. Gary Wittlich. Englewood Cliffs, NJ : Prentice-Hall.

Wolpe, Stefan

1967 Thinking twice. In Schwartz and Childs (1967:274-307).

1979 Thoughts on pitch and some considerations connected with it. *Perspectives of New Music* 17/2:28-57.

1983 Any bunch of notes. Austin Clarkson, ed. *Perspectives of New Music* 21/1-2:295-310.

1984 On new (and not-so-new) music in America. Trans. Austin Clarkson. *Journal of Music Theory* 28/1:1-45.

1986 Lecture on Dada. Austin Clarkson, ed. *Musical Quarterly* 72/2:202-215.

Wuorinen, Charles

1979 *Simple composition*. New York: Longman.

Xenakis, Iannis

1971 *Formalized music: Theory and method in composition.* Bloomington: Indiana University Press.

1985 *Arts, sciences: alloys.* Trans. Sharon Kanach. New York: Pendragon Press.

Yasser, Joseph

1932 *A theory of evolving tonality*. Reprinted in New York: Da Capo Press, 1975.

Zhang, Fan

1993 *Degree sequence polyhedra and separation algorithms*. Ottawa: National Library of Canada.

Zhang, Limin

1985 Every 4-regular simple graph contains a 3-Regular subgraph. *Journal of Changsha Railway Institute* 1:130-154.

About the Author

JEFFREY JOHNSON is currently Associate Director of Artistic Education and Chair of the Music Theory Department at the renowned Boys Choir of Harlem. He has degrees in musical composition from Boston University, Eastman School of Music, and Ithaca College. He is an active composer, conductor, and theorist. His first book *Thesaurus of Abstract Musical Properties* was published by Greenwood Press in 1995.

www.ingramcontent.com/pod-product-compliance
Ingram Content Group UK Ltd.
Pitfield, Milton Keynes, MK11 3LW, UK
UKHW041637031026
15587UKWH00001B/346

9 780313 301582